T0395419

Published by Princeton University Press
41 William Street, Princeton, New Jersey 08540
99 Banbury Road, Oxford OX2 6JX
press.princeton.edu

Library of Congress Control Number 2024944450

ISBN 978-0-691-26248-2
Ebook ISBN 978-0-691-26633-6

Typeset in Trajan and Caslon

Printed and bound in Malaysia
1 3 5 7 9 10 8 6 4 2

British Library Cataloging-in-Publication
Data is available

This book was conceived, designed, and produced by
UniPress Books Limited
Publisher: Jason Hook
Managing editor: Slav Todorov
Project manager: Katie Crous
Art direction: Alexandre Coco
Design: Luke Herriott
Illustrator: Claudia Myatt
Diagrams: Rob Brandt
Picture researcher: Elaine Willis

Cover design: Luke Herriott
Cover images: (Front from top to bottom): *Cloud Study*, Knud Baade, 1838; *Cloud Study* (*Other*), Knud Baade, 1838; *Cloud Study Over Landscape*, Knud Baade, 1838; *Cloud Study*, Knud Baade, 1852; *Cloud Study (Other)*, Knud Baade, 1838; *Cloud Study*, Knud Baade, 1850; *Cloud Study*, Johan Christian Dahl, 1843. Courtesy of Nasjonalmuseet for Kunst, arkitektur og Design / The Fine Art Collections. (Back): iStock

CLOUDS

HOW TO IDENTIFY NATURE'S MOST FLEETING FORMS

EDWARD GRAHAM

PRINCETON UNIVERSITY PRESS
PRINCETON AND OXFORD

CONTENTS

FOREWORD

"THE SERVICE OF CLOUDS": ART, SCIENCE, AND THE SKY

By Richard Hamblyn

When the British painter Joseph Wright (of Derby) climbed Mount Vesuvius during a visit to Naples in 1774, he said that he wished he could have made the ascent in the company of a geologist, since "his thoughts would have center'd in the bowels of the mountain; mine skimmed over the surface only. There was a very considerable eruption at the time, of which I am going to make a picture."

An erupting volcano was nothing new in late eighteenth-century art, but what was new was Wright's acknowledgment of his scientific limitations, the artistic implications of which were clear: if witnessing a natural spectacle was one thing, but trying to understand it was another, wouldn't knowing the science of what one was seeing make one a better artist? This was a profound and troubling question—does perception depend on understanding?—and over the course of the next half-century, amid the continued rise of Enlightenment natural philosophy, it became the question that every serious landscape painter found themselves working to answer.

Artists had, of course, depicted land, sea, and sky for centuries, with varying degrees of success. By the seventeenth century, painting treatises offered detailed instructions for sketching trees and clouds, encouraging artists to get out into nature and paint in the open air. The Dutch marine painter Willem van de Velde the Younger made a series of plein air cloud sketches on Hampstead Heath during his time in London in the 1680s, a process which he called "going a skoying," while later treatises, such as Alexander Cozens' *New Method of Assisting the Invention in Drawing Original Compositions of Landscape* (1785), offered graduated visual classifications of sky-states, from "all plain" to "all cloudy," via every shade of cloudiness in between. But none of this guidance required a painter to know anything about the mechanics of what they were painting, and for most of art history the sky was viewed as little more than a moving canvas to be copied.

But that was set to change, and by the time the critic and artist John Ruskin declared, in 1844, that "every class of rock,

1.

earth, and cloud, must be known by the painter, with geologic and meteorologic accuracy," the practice of landscape painting had undergone a transformation into something akin to natural history fieldwork. This transformation was exemplified by John Constable's remarkable claim, made during a lecture in 1836, that "painting is a science and should be pursued as an inquiry into the laws of nature. Why, then, may not landscape painting be considered a branch of natural philosophy, of which pictures are but the experiments?"

Constable's alignment of painting and natural philosophy arose in the context of large-scale cultural changes that had led to the emergence of science and technology as the economic and intellectual driving forces of much of the Western world. The rapid professionalization of the sciences, beginning in the late eighteenth century, saw an array of specialist scientific bodies established over the course of a couple of decades, including (in London alone) the Linnean Society (founded 1788, for the study of botany and taxonomy), the Geological Society (1807), the Astronomical Society (1820), and the Meteorological Society (1823), of which Ruskin was an early member—in fact, one of his first publications, "Remarks on the Present State of Meteorological Science," appeared in the inaugural volume of the Society's Transactions in 1839, when Ruskin was just twenty years old.

1. ***Vesuvius from Posillipo by Moonlight*** **by Joseph Wright of Derby, ca. 1788**
Early cloud paintings commonly depict *Cumulus congestus* and *Cumulonimbus* as towering turrets that lend a dramatic majesty to the amphitheater of the skies. On this occasion, the artist could be justified, as the Vulcan god may be providing enough heat and moisture to create and sustain a powerful *Cumulonimbus calvus flammagenitus* (page 194).

2. ***Cloud Study*** **by John Constable, 1822 (overleaf)**
Towering *Cumulus congestus* rise majestically upward, indicative of rapidly ascending moist updrafts of air. The hazy atmosphere, a characteristic of Constable paintings, helps to moderate the contrast between the dark shadows and the clouds' highly reflective upper surfaces. The atmosphere is ripe for a thunderstorm.

2.

So, when Constable ventured onto Hampstead Heath in the early 1820s, with the aim of painting clouds in situ, it may have been under the same gray skies as his Dutch predecessor, van de Velde the Younger, but it was in a wholly different conceptual world. He took with him a copy of Thomas Forster's *Researches About Atmospheric Phænomena* (1815), the first chapter of which offered an illustrated summary of Luke Howard's recent classification and nomenclature of clouds. Howard, like many of his scientific contemporaries, was an accomplished draftsman, and his landmark *Essay on the Modifications of Clouds* (1803), as did Forster's later summary, featured engravings of his own watercolor studies of the seven cloud types that he had identified and named: *Cirrus*, *Cumulus*, *Stratus*, and their compounds. Constable's penciled annotations to Forster's textbook confirm his up-to-date knowledge of clouds and weather, as do the detailed weather-notes that he added to the more than one hundred oil-on-paper cloud sketches he completed between 1820 and 1822, which are now among his most celebrated works.

Constable was not the first Romantic-era artist to engage with Luke Howard's nephological insights. When a German translation of Howard's cloud essay was read by the polymath J. W. von Goethe, he sent copies to several artists of his acquaintance, advising them to study it before going out to sketch in the open air. For the Dresden-based painter Carl Gustav Carus, the advice proved revelatory: as soon as he read Howard's essay, he wrote, he felt that the "problem of how to reconcile scientific analysis with creative freedom had now been solved," since clouds, according to Howard's new system, had become explicable while remaining free to go about their ceaseless transformations. His response was widely shared, and a generation of predominantly northern European landscape painters, including the Danish C. W. Eckersberg, the Danish–Norwegian Johan Christian Dahl, his compatriot Knud Baade, and the Dutchman Anton Pitloo, were won over to this new way of thinking about the sky and its place in art.

Many, but not all: In 1817 Goethe had approached Caspar David Friedrich with a request for a set of illustrations for the German translation of Howard's essay, but Friedrich declined on the grounds that it would "undermine the whole foundation of landscape painting," with the artist declaring himself opposed to "any attempt to force the free and airy clouds into a rigid order and classification." Clouds, for High Romantics such as Friedrich, remained symbols of elemental freedom, although the work of subsequent generations of artists, such as the Norwegian Lars Hertervig or the American Frederic Edwin Church, would be further shaped by the natural sciences, with Church making dozens of annotated cloud studies in preparation for some of his best-known canvases.

3.

As with Constable's clouds a generation before, these plein-air sketches started off as exercises in rapid execution—done in pursuit of technique rather than transcendence—but, as can be seen on every page of this beautifully illustrated book, something about their weightless mutability speaks to our modern sensibilities more directly than many of these painters' more polished studio productions. And though the advent of photography in the late nineteenth century would reshape the contours of both the arts and the sciences—especially the more observational sciences such as meteorology—Ruskin's claim, made in 1856, still holds beguilingly true:

"...if a general and characteristic name were needed for modern landscape art, none better could be invented than 'the service of clouds.'"

3. ***Looking Across the Hudson Valley, New York*** **by Frederic Edwin Church, ca. 1867**
The Hudson River School was a nineteenth-century art movement known for idealized depictions of American landscapes. In this masterpiece, Church captures the last rays of the setting Sun as it illuminates a broken deck of mid-level *Altocumulus*. Below lies a darkened bank of lumpy *Stratocumulus*. The pale green light emanating from behind the clouds epitomizes perfectly an airmass of northern or polar origin.

4. ***The Valley of the Seine at Saint-Cloud*** **by Alfred Sisley, 1875 (overleaf)**
Fair-weather cumulus arranged in "cloud streets" (*Cumulus humilis radiatus*), on a fine day over the appropriately named Saint-Cloud, France.

THE CLOUD CLASSIFICATION TABLE

	GENERA (TYPE)	SPECIES (CAN BE ONLY ONE)	VARIETIES (CAN HAVE MORE THAN ONE)	SUPPLEMENTARY FEATURES	ACCESSORY CLOUDS
	CIRRUS	*fibratus* *uncinus* *spissatus* *castellanus* *floccus*	*intortus* *radiatus* *vertebratus* *duplicatus*	*mamma* *fluctus*	
	CIRROCUMULUS	*stratiformis* *lenticularis* *castellanus* *floccus*	*undulatus* *lacunosus*	*virga* *mamma* *cavum*	
	CIRROSTRATUS	*fibratus* *nebulosus*	*duplicatus* *undulatus*		
	ALTOCUMULUS	*stratiformis* *lenticularis* *castellanus* *floccus* *volutus*	*translucidus* *perlucidus* *opacus* *duplicatus* *undulatus* *radiatus* *lacunosus*	*virga* *mamma* *cavum* *fluctus* *asperitas*	
	ALTOSTRATUS		*translucidus* *opacus* *duplicatus* *undulatus* *radiatus*	*virga* *praecipitatio* *mamma*	*pannus*

	GENERA (TYPE)	SPECIES (CAN BE ONLY ONE)	VARIETIES (CAN HAVE MORE THAN ONE)	SUPPLEMENTARY FEATURES	ACCESSORY CLOUDS
	NIMBOSTRATUS			*praecipitatio* *virga*	*pannus*
	STRATOCUMULUS	*stratiformis* *lenticularis* *castellanus* *floccus* *volutus*	*translucidus* *perlucidus* *opacus* *duplicatus* *undulatus* *radiatus* *lacunosus*	*virga* *mamma* *praecipitatio* *fluctus* *asperitas* *cavum*	
	STRATUS	*nebulosus* *fractus*	*opacus* *translucidus* *undulatus*	*praecipitatio* *fluctus*	
	CUMULUS	*humilis* *mediocris* *congestus* *fractus*	*radiatus*	*virga* *praecipitatio* *arcus* *fluctus* *tuba*	*pileus* *velum* *pannus*
	CUMULONIMBUS	*calvus* *capillatus*		*praecipitatio* *virga* *incus* *mamma* *arcus* *murus* *cauda* *tuba*	*pannus* *pileus* *velum* *flumen*

1.

INTRODUCTION

We live at the bottom of an ocean—a sea of fluid, which is both above us and all around us. This is what we call the atmosphere, or "air." Fortunately, it is a largely invisible fluid that is neither too compressed nor dense for us to live and breathe in, at least at ground level. And there is one special, albeit very tiny, constituent of this gaseous mix that dictates the appearance of our atmosphere: water vapor.

It is water vapor alone that rises and condenses into the clouds we know so well, bringing us our weather, life-giving rains, "fresh showers for the thirsting flowers" (from the poem "The Cloud" by Percy Bysshe Shelley)—and crops. It is water vapor alone that physicists tell us has the highest known "latent heat" of any substance, a kind of secret heat energy that is released upon its evaporation, and which is returned to the air upon condensation as a cloud, giving it an extra buoyant boost. And it is the water vapor molecule alone that has a molecular mass less than two-thirds that of its much more abundant atmospheric neighbors, nitrogen and oxygen—meaning that moist air rising into clouds really is "lighter than air."

It is no wonder that clouds have intrigued us since time immemorial. For both scientists and artists, it is surely their esthetic beauty, and the emotional response they engender deep within us, that can be regarded as the main factors in motivating our interest in them. As clouds are in a continuous state of modification, or evolution, an art form lying somewhere close to the Impressionist movement of the late nineteenth century is perhaps the best way to capture and understand them. That idea underpins the approach of this book, which combines modern meteorology with cloud studies by some of the greatest artists ever to look skyward.

Clouds are very much more than "airy nothings" (as described by Shakespeare in *A Midsummer Night's Dream*). In our digital age of fleeting online "stories," the skyscape can be considered a unique, active, 24/7, never-to-be-repeated, real-world livestream, demonstrating the unerring laws of physics of the atmosphere. As we continue to tinker directly with the composition of air, through our pollution and the addition of greenhouse gases, the clouds and extreme weather that we are newly experiencing are direct manifestations of our behavior. Like it or not, we are now the cloud makers.

1. ***Cloud Study*** **by Frederic Edwin Church, ca. 1868–69**
Church captures the evening light perfectly in this study. A "street" of *Cumulus mediocris radiatus*, aided by a light breeze blowing from left to right, is backdropped against the pale blue of a humid summer atmosphere. Patches of *Cumulus fractus* or *Stratocumulus* nearby may indicate the remains of earlier clouds. In the background, high in the atmosphere, are streaks of *Cirrus* or *Cirrostratus fibratus*, tinged a pinky-ocher by the setting Sun.

KNUD BAADE (1808–79)

Strongly influenced by Caspar David Friedrich, Baade was a Norwegian painter who was especially fond of moonlit landscapes and dramatic use of chiaroscuro. He suffered from ill health throughout his life yet traveled throughout Europe in search of the most vivid landscapes, which often featured especially beautiful cloud formations. In later life, he returned to his native Norway, where he created some of his most impressive studies of the unique Norwegian coast.

2. ***Cloud Study* by Knud Baade, 1838**
Here, Baade has acutely and realistically captured the low-Sun and reduced-light conditions of northern winter skies close to twilight. The atmosphere is unstable, yielding great turrets and towers of *Cumulus congestus* and nascent *Cumulonimbus*, in a coastal airmass of polar origin, attested to by the pale green background and Arctic blue skies.

The juxtaposition within the same frame of both towering *Cumulus* and more stable clouds (the spreading out of *Stratocumulus cumulogenitus* in the foreground, or is it the misplaced anvil of a *Cumulonimbus*?) suggest that the artist's primary aim on this occasion was to achieve an atmosphere of emotional melodrama, rather than an exact meteorological snapshot.

6

5

4

3

2

1

"As clouds ascend,
are folded, scatter, fall,
Let the world think of thee
who taught it all."

Johann von Goethe, *In Honour of Howard* (1821)

CLASSIFYING CLOUDS

THE 10 PRINCIPAL CLOUD TYPES

The World Meteorological Organization's (WMO) *International Cloud Atlas* provides an internationally agreed standard for the observation and reporting of cloud types. As first proposed by Luke Howard in his *Essay on the Modifications of Clouds* in 1803, and later adopted by the WMO, it uses a Linnaean taxonomic system, similar to the hierarchical way that plants and animals are scientifically named.

The most recent version of the atlas, which was published online in 2017, lists ten principal cloud genera (types), as shown in the Cloud Diagram on the page opposite. These genera can be further sub-divided into fifteen unique species—only one species is possible per cloud, and *Altostratus* and *Nimbostratus* have no sub-species. An additional nine cloud varieties, eleven supplementary features, and four cloud accessories are also possible: each cloud can have one, more than one, or none of these. However, cloud varieties, accessories, and supplementary features tend to be the exception rather than the rule, and many are exclusive to particular cloud genera only—they are listed in full in the Cloud Classification Table on pages 14–15.

The Cloud Diagram and Cloud Classification Table only depict clouds of the troposphere—the region of the atmosphere where our everyday weather occurs. All tropospheric clouds are the result of, and can forewarn us of, the weather conditions that we experience on Earth's surface on any given day. Occasionally, other types of clouds may appear in the stratosphere and mesosphere, some of which we will discuss later in this book. These are much more tenuous, ethereal, and other-worldly than those of the troposphere, and do not directly impact everyday weather on the ground.

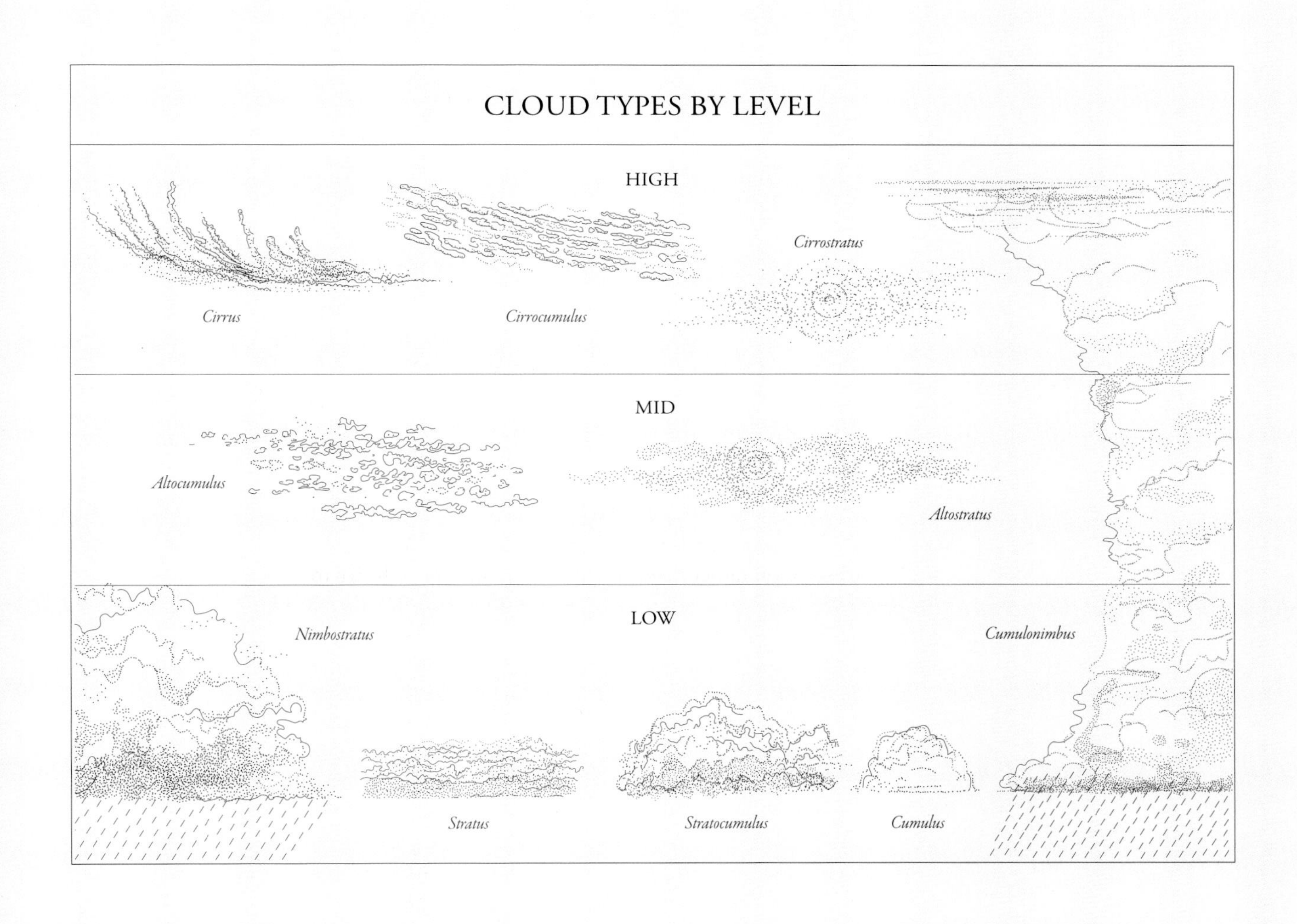
CLOUD TYPES BY LEVEL
HIGH
Cirrus
Cirrocumulus
Cirrostratus
MID
Altocumulus
Altostratus
LOW
Nimbostratus
Cumulonimbus
Stratus
Stratocumulus
Cumulus

EARTH'S ATMOSPHERE	
Gas	**Total % by mass of dry air**
Nitrogen	78.08
Oxygen	20.95
Argon	0.93
Carbon dioxide	0.04
Water vapor (global mean)	0.25

DALTON'S LAW

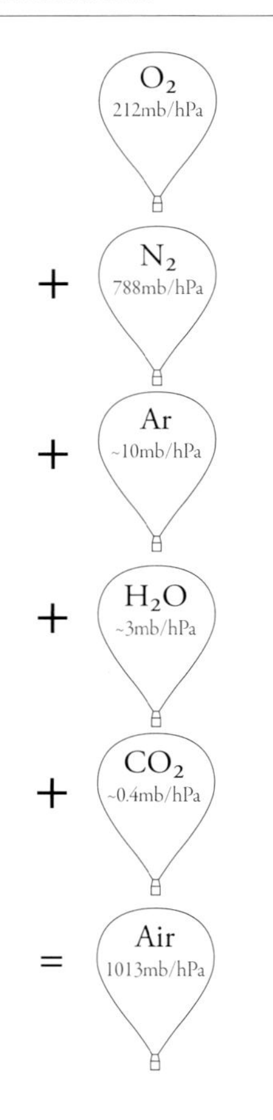

Dalton's Law
Schematic of Dalton's Law, which states that the total atmospheric pressure is equal to the sum of the partial pressures of the individual gases.

ORIGIN OF THE ATMOSPHERE AND CLOUDS

The atmosphere is composed of gases that have failed to escape the gravitational force of Earth, and it has evolved continuously during the history of our planet. On the early, molten proto-Earth (perhaps a bit like Saturn's moon, Titan, today), some of the first constituents of the atmosphere were methane, carbon dioxide, and hydrogen sulfide. Liquid water did not condense to form the oceans until much later when the planet cooled.

Oxygen only began to appear after one billion years of Earth's history, when simple-celled life in the world's seas started to photosynthesize, using up carbon dioxide in the presence of sunlight; it took more than 90 percent of Earth's known history before the atmosphere began to contain anything close to today's amount of oxygen. Our familiar blue skies did not materialize until after another billion years had passed by.

Tiny details matter

In today's atmosphere, there is approximately 78% nitrogen, 21% oxygen, and 1% argon, with some trace amounts of water vapor (0.25%) and carbon dioxide (0.04% but increasing). It is rather curious that those final two constituents, water vapor and carbon dioxide, are so vital in producing the weather, climate, and indeed life on Earth as we know it, despite being present in such tiny concentrations. Without water vapor, there would be no clouds, no rain, and no life! And even if there was life, without carbon dioxide we would be entombed in a quasi-permanent ice age.

It is water vapor that rises to condense and form the clouds we see, and we readily acknowledge its fundamental importance along with that of oxygen. It is perhaps too easy to dismiss the influence of nitrogen and argon, but the simple fact is that all the gases play a role in forming the weather and clouds with which we are so familiar. This is because they all exert a partial pressure, which when summed together gives the total pressure at the surface of Earth according to Dalton's Law, and their combination is what we call "air."

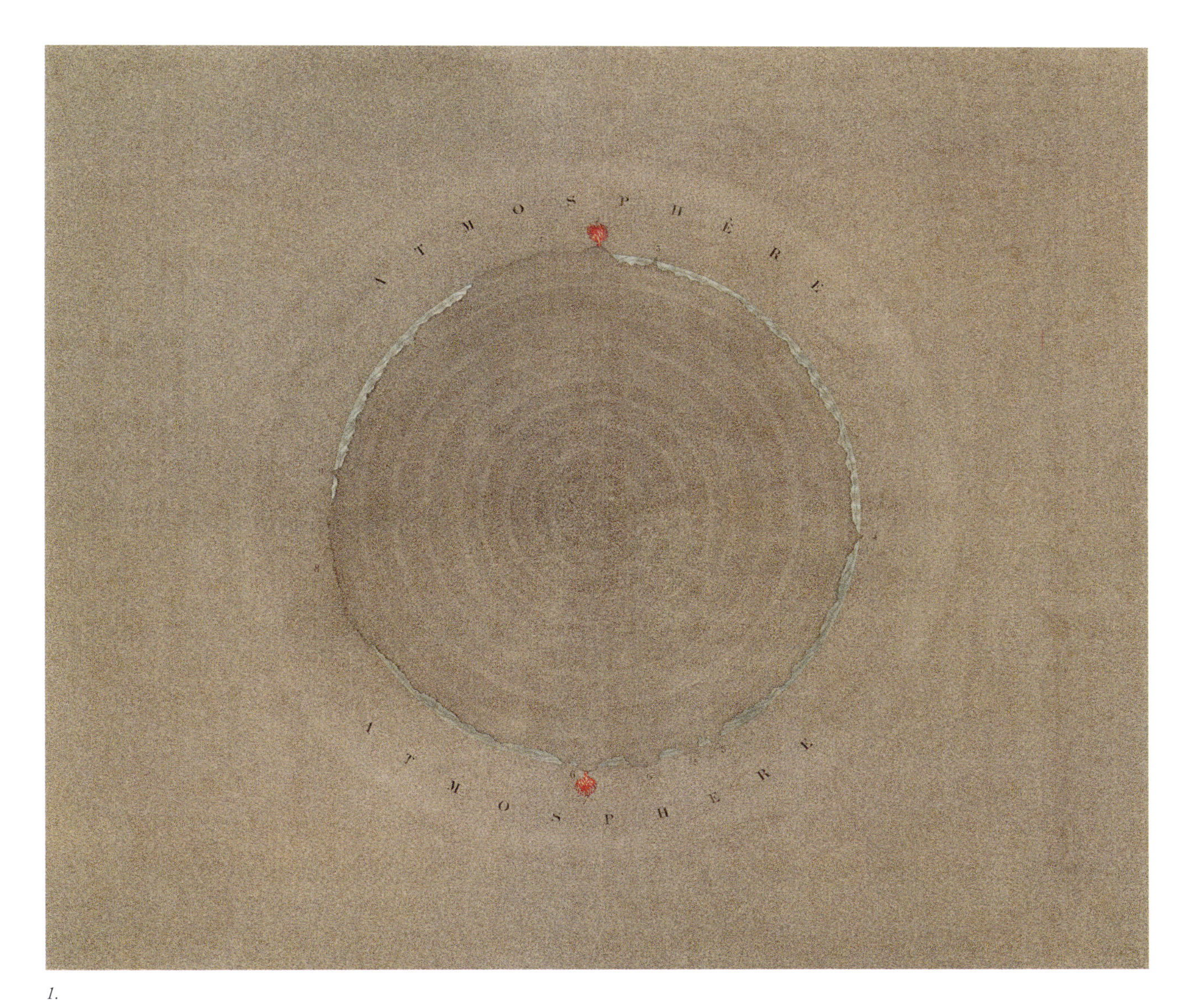

1.

1. ***Earth and its Atmosphere* by Sigismond Visconti, 1839**
   This is perhaps the first representation of Earth's atmosphere as the "Thin Blue Line" (as it is described today by NASA scientists), shown as a cross-section with radiating bands of shading to illustrate the planet's layers.

THE ATMOSPHERE

The atmosphere is composed of many layers of air of slightly differing composition but of enormously differing air pressure, namely the troposphere, stratosphere, mesosphere, and thermosphere. Each layer is separated from its neighbors by a change in lapse rate (the rate of change of air temperature with height), which controls the stability of the layers and largely prevents them from mixing with one another. Air is highly compressible, so air pressure decreases extremely rapidly with height—the relationship is exponential. As a result, more than half the mass of the atmosphere lies below an altitude of 3½ miles (5.5 km), and 99 percent of it lies below 18½ miles (30 km). This is why climbers of the world's tallest mountains must take oxygen supplies with them to survive for any length of time at high altitude.

When it comes to clouds, it is the troposphere we are mainly interested in, because all our weather takes place within this very thin layer bordering Earth's surface; it is only 5–11 miles (8–18 km) thick, depending on where you are in the world. It is also here—but only within the lowest mile or two—that humans have evolved to live, breathe, work, and die.

The troposphere effectively contains all Earth's atmospheric water vapor, most of which is to be found in a highly concentrated thin layer close to the ground. The vast majority is located in the tropics: here it can be up to a maximum of 4 percent by mass, whereas it accounts for only 0.25 percent on average for the whole atmosphere. Interestingly, the amount of water vapor that air can hold is controlled solely by air temperature (again, it is an exponential relationship), and does not depend on air pressure. This uncompromising dependence of water vapor on temperature means that rising air currents will almost inevitably saturate to form clouds in the troposphere as soon as they become cold enough.

At the top of the troposphere lies the "tropopause," which is the boundary with the overlying stratosphere. Here, a strong temperature inversion (a reversal of the usual decrease of temperature with height) prevents the exchange of air between the two layers, for the most part. The actual height of the tropopause varies somewhat from season to season, and between the polar regions and the tropics.

As air is invisible, we cannot see these layers individually. That said, within the lowest layer—the troposphere—clouds act as fairly good tracers of atmospheric flow and are usually restricted in penetrating beyond the tropopause.

Earth's atmosphere
Schematic of Earth's atmosphere indicating the altitude of its various layers, including their approximate temperature and pressure. The vertical axis is not to scale.

EARTH'S ATMOSPHERE

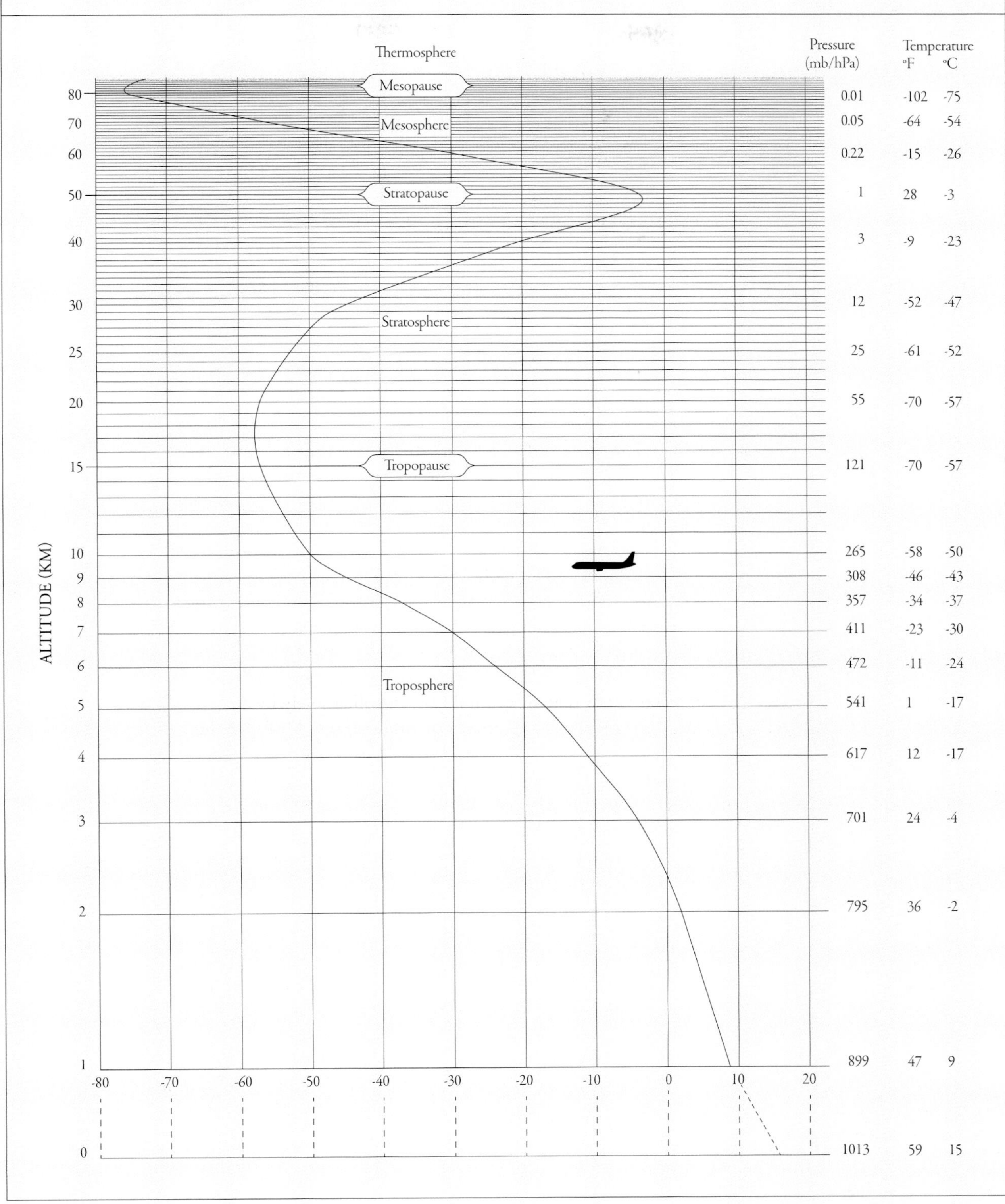

CLASSIFICATION OF CLOUDS BY HEIGHT

Today, the World Meteorological Organization (WMO) classifies the altitude of clouds, albeit a little arbitrarily, into three main categories: low-level, mid-level, and high-level clouds. These levels refer only to the height of the base of the cloud above ground level, so some vertically thick clouds—for example *Nimbostratus* and *Cumulonimbus*—may extend across two levels, or indeed all three. Nevertheless, they remain classified by their lowest level.

It is worth noting that the WMO classifications refer to the relative height of clouds above the surrounding ground level, not their absolute height above sea level. Low-level *Cumulus* clouds that form over high mountain ranges, or over upland plains such as the Colorado or Tibetan plateaus, therefore remain classified as low-level clouds, despite the fact that they may be forming at the same altitude, relative to sea level, as mid-level cloud species over the oceans.

In addition, due to the colder—and therefore denser—troposphere over polar regions, the altitudinal ranges used to define the low-, mid-, and high-level cloud categories vary considerably between the tropics, subtropics, middle latitudes (temperate regions), and high latitudes (polar, Arctic, and Antarctic regions). This, in part, reflects the fact that in tropical climates the freezing (or glaciation) level of clouds in the atmosphere is very high—more than 3–4 miles (5–6 km)—but this gets lower as we get closer to the poles.

HEIGHT RANGES

Height ranges (above ground level) of low-, mid-, and high-level clouds, according to WMO.

LEVEL/REGION	
High level	
Midlevel	
Low level	

THE NAMING OF CLOUDS

2. ***Portrait of Luke Howard* by John Opie**
Ordering and classification were important aspects of Enlightenment science. Fascinated since childhood by the weather, and clouds in particular, Howard classified and named different cloud types between 1803 and 1811, providing sketches and illustrations.

3. **Cloud studies by Luke Howard, ca. 1808–11**
Here we have an idealized, almost schematic, depiction of *Cumulus mediocris* and *Cumulus congestus*, with characteristic flat bases, building vertically upward into a shallow *Cumulonimbus capillatus*, sporting a symmetrical flat-top "anvil" (*incus*). At upper left, there is a patch of *Cirrostratus cumulonimbogenitus*.

4. Large turrets of *Cumulus congestus* / *Cumulonimbus calvus praecipitatio* pierce through multiple layers of *Stratus cumulogenitus* or *Altostratus cumulonimbogenitus*.

Luke Howard was born into a well-to-do Quaker family in London, England, in 1772. A quiet, modest, unassuming gentleman, he became a pharmacist by profession, operating his own chemist shop on Fleet Street, and later managing a pharmaceutical factory in east London.

While attending a strict school where the learning of Latin took precedence over other subjects, he developed a passion for observing the natural environment, setting up a small weather station in the garden of his parents' home to record the temperature, precipitation, and air pressure. He also maintained a strong interest in botany.

Along with other Protestant Dissenters in the early nineteenth century, Quakers were still excluded from attending English universities or from joining business guilds. Having to look elsewhere for professional stimulation, Howard joined the Askesian Society, a scientific debating club. In December 1802, he presented his paper *On the Modifications of Clouds* before the Society, and this was later published as a series of essays in the *Philosophical Magazine*. The manuscripts were later expanded into booklet form, which was translated into French and German.

With increasing acclaim, Howard continued his meteorological research. In 1821, he was elected as a Fellow of the Royal Society. Later, he co-founded the Meteorological Society of London, the precursor of the Royal Meteorological Society. In 1837, he published *Seven Lectures in Meteorology*, one of the first meteorological textbooks. After spending many of his later years living in Yorkshire, Howard returned to London in 1852, where he lived to the ripe old age of 91.

The fundamental simplicity of Howard's scheme lay in the intuitive realization that although clouds have many shapes, in fact an infinite number, they have just a few basic forms—or three "simple modifications," according to Howard. Furthermore, and this was the crux of the matter, the three basic forms could mutate and "pass into another," forming either two "intermediate modifications" or two "compound modifications." In total, this yielded seven different cloud genera, as listed in the table on page 33. These can be compared with today's WMO *International*

2.

3.

4.

5.

6.

Cloud Atlas (page 22), which comprises ten different cloud genera, and includes most of Howard's original system.

Howard was not alone in his attempts to name the clouds. Around the same time, and completely unbeknownst to him, Jean-Baptiste Lamarck, a zoologist and member of the French Academy of Sciences, had proposed five cloud categories that were similar to Howard's, but containing French names. Lamarck also suggested that clouds might be categorized by altitude, an attractive idea that would later be adopted by the International Meteorological Organization (the forerunner of the WMO) in 1896. However, it would be Howard's Linnean system of Latin cloud names that was to catch on, probably because of the universality of Latin and its widespread use in the classification of the biological sciences, but also due to its very simple, yet elegant, scheme of cloud "modifications" that Howard had uniquely devised. So, Howard's school Latin did finally come in useful!

5. ***Aggregate Cumulus in Different Stages* by Luke Howard, ca. 1803–11**
Cumulus fractus (left), *Cumulus humilis* (center), *Cumulus mediocris* (right), *Cumulus congestus* (rear), with a small patch of *Stratus cumulogenitus* (left of center).

6. ***Bank of Cumulus Lit from Behind by Sun* by Luke Howard, ca. 1803–11**
Here, Howard was demonstrating both the form (shape) of *Cumulus*, as well as how the water-droplet-rich clouds can appear both dazzling white and dark and threatening, depending on the depth of cloud and the position of the Sun in relation to the cloud and the observer.

Howard's 1803 cloud "Modifications" and the WMO equivalent cloud genus and level.

CLOUD MODIFICATIONS	
1803 CLOUD MODIFICATIONS	PRESENT WMO* GENUS AND LEVEL
Cirrus	*Cirrus* (high)
Cumulus	*Cumulus* (low)
Stratus	*Stratus* (low)
INTERMEDIATE MODIFICATIONS	
Cirro-Cumulus	*Cirrocumulus* (high)
Cirro-Stratus	*Cirrostratus* (high)
COMPOUND MODIFICATIONS	
Cumulo-Stratus	*Stratocumulus* (low)
Cumulo-Cirro-stratus or *Nimbus*	*Nimbostratus* (low) and *Cumulonimbus* (low)

* *International Cloud Atlas* (2017), WMO

THE *INTERNATIONAL CLOUD ATLAS*

First published in 1896, the *International Cloud Atlas* provides an agreed worldwide standard for the observation and reporting of cloud types. Having an agreed standard was a vital step in the development of meteorology as a science, because if there were to be any hope of successful weather forecasts, information needed to be shared quickly between countries using an established language.

Since its inaugural publication in 1896, the *International Cloud Atlas* has undergone occasional revisions, conducted under the auspices of the WMO, based in Geneva. Despite more than two centuries having elapsed since its publication, Howard's original Linnean nomenclature has endured, although changes and additions have occurred. For example, in 1870, Émilien Renou, director of the Saint-Maur-des-Fossés observatory in France, proposed the introduction of both *Altocumulus* and *Altostratus*, reflecting their separate mid-level status, lying above the low-level clouds of *Stratus* and *Cumulus* but below high-level *Cirrus*. *Cumulonimbus* was proposed as a separate cloud by French meteorologist Philip Weilbach in 1880. Cloud definitions based around low, mid-, and high cloud levels, as originally suggested by Lamarck back in 1802, were finally adopted at the International Meteorological Congress of 1896. In 1930, the International Commission for the Study of Clouds accepted *Nimbostratus* as another distinct genus.

The atlas underwent the greatest number of changes in its history in its most recent version, published online in 2017, when twelve new variety, supplementary, and accessory cloud names were adopted. This was a reflection of the rapid emergence of smartphone and digital photography technology, as well as citizen science initiatives during the first two decades of the twenty-first century, which provided clear and unambiguous evidence of some rare, unusual, and local clouds that were hitherto little known, in addition to clouds relating to human activity. The total revisions comprised one new cloud species (*volutus*, page 198), five new supplementary cloud features (*asperitas*, *cavum*, *murus*, *cauda*, and *fluctus*, page 198), one new accessory cloud type (*flumen*), and five new special "mother" clouds (*cataractagenitus*, *flammagenitus*, *homogenitus*, *homomutatus*, and *silvagenitus*, page 194).

In total, the current *International Cloud Atlas* contains 10 separate cloud genera, 15 species, 9 varieties, 11 supplementary, and 4 accessory clouds. The species, varieties, supplementary, and accessory clouds can be thought of simply as fresh additional Howard "modifications."

7. ***Nimbus*, from the *International Cloud Atlas*, 1896**
   Today, this would probably be classified as *Nimbostratus virga pannus*.

8. ***Stratus*, from the *International Cloud Atlas*, 1896**
   Although it captures the monotone, featureless, gray expression of *Stratus* very well, the lens-shaped holes in the cloud, with blue sky beyond, clearly indicate a influence of mountain waves, so today the cloud would be more strictly classified as *Stratocumulus lenticularis*.

7.

8.

CLOUD SYMBOLS

For a weather report to be of use to anyone, information on cloud development and evolution must be transferred *more quickly than the speed of propagation of the weather itself.* The information also needs to be communicated quickly in a language or code that is both easily composed and easily deciphered. This did not become routine nor did the cost become feasible until after the invention of the telegraph in the mid-1830s.

In North America, the Smithsonian Institution helped to establish the first weather network reporting by telegraph in 1849. By 1860, the network had grown to 500 daily reporting stations.

In Great Britain and Ireland, Admiral Francis Beaufort of the Royal Navy first developed a standardized scale for the reporting of wind speed—the Beaufort Scale—as well as the collection of meteorological data in shortened, coded formats. In 1854, Vice-Admiral Robert Fitzroy was appointed to lead a new department to collect meteorological data, which was later to develop into the Met Office. Following a naval tragedy in 1859, Fitzroy expanded this service into short-term predictions using an analog method based around the extrapolation of the movement of weather systems and their common patterns. Fitzroy coined the term "weather forecast" to describe his attempts, publishing these from 1861 onward in *The Times*.

Even before the Linnean system for the nomenclature of clouds was proposed by Luke Howard in 1803 (page 30), scientists had realized that a simple, concise but internationally agreed system of weather and cloud reporting was necessary. In 1771, Johann Heinrich Lambert proposed a set of elementary symbols to describe cloudy skies, fog, precipitation, or thunder. Later, Howard, in his 1803 treatise, proposed a simple set of dashes, strokes, and semicircles to depict the principal cloud types, echoes of which can be found in today's internationally-accepted WMO's *Manual of Codes*. For many decades throughout the twentieth century, and until well after the dawn of the computer era, an unspoken generation of operational meteorologists would laboriously plot these symbols by hand on large synoptic charts, sometimes every hour, day and night, after deciphering the telegraph and teletext codes for each available weather station. Today, this task is mostly computerized.

WMO cloud abbreviations and their respective symbols of each of the ten different cloud genera.

WMO CLOUD ABBREVIATIONS		
SPECIES	ABBREVIATION	SYMBOL
Cirrus	Ci	
Cirrocumulus	Cc	
Cirrostratus	Cs	
Altocumulus	Ac	
Altostratus	As	
Nimbostratus	Ns	
Stratocumulus	Sc	
Stratus	St	
Cumulus	Cu	
Cumulonimbus	Cb	

JOHN RUSKIN (1819–1900)

Ruskin was a vocal critic of industrialization, believing it to be responsible for the moral, social, and environmental degradation of his age. He abhorred the pollution created by Victorian factories and we may even regard him as an early advocate for action to combat the climate crisis. His painting *Chasing the Storm Cloud* is an expression of the menacing and unnatural weather patterns Ruskin attributed to climate change, which he feared would alter the landscape of England forever.

CLOUD GEOMETRY

Putting Pythagoras and the protractor aside for a moment, let us consider a rather loose, creative meaning of the term "geometry"—that is, the shape and relative arrangement of things, even when they are in a fluid state and undergoing constant change. We can now perhaps use the word to describe the various ephemeral forms and structure of clouds as they continuously grow, develop, and decay before our eyes.

Most clouds grow by two processes: either by convection or by gentle uplift. Both processes cause cooling of air and, if the air is moist enough, its saturation, leading to the formation of a cloud. Convection is the same process as buoyancy and refers to the rapid rise of moist air bubbles in the atmosphere, caused by heating from below. Although each individual thermal rises in a turbulent and somewhat chaotic manner, in well-mixed air masses moving over smooth terrain, these local idiosyncrasies get averaged out, resulting in the development of regular patterns in both the horizontal and vertical, such as in the regular spacing of *Cumulus* cloud streets (page 96), the fractal-repetitive nature of their ascending *Cumulus* turrets, or their flat, level bases (page 98).

In contrast, gentle uplift takes place at boundaries of air masses, for example along a weather front where cold, dry, dense air is overridden by an advancing warm, moist, less dense air mass, causing condensation and, usually, layer, or stratiform, clouds. When rising air is unable to ascend any more, it may descend again, or more often it spreads sideways, forming widespread layers of cloud. Changes in temperature, humidity, wind speed, and wind direction with height, as well as the presence of obstructions such as mountains, impact the flow of air in these circumstances, and may result in regular patterns, such as billow clouds (page 154) or mountain lee wave clouds (pages 158–9).

The geometric patterns and similarities seen in clouds are just another example of the regular patterns that frequently appear across the natural world, for example in ammonite shells, the leaf of a fern, a zebra's stripes, or the hexagonal columns of basalt at the Giant's Causeway in Ireland. At first sight, all of these seem to act contrary to the general direction of the entropy (a measure of the amount of energy unavailable to do work) of the universe, which both cosmologists and the laws of thermodynamics say is pushing us toward a continually greater state of disorder.

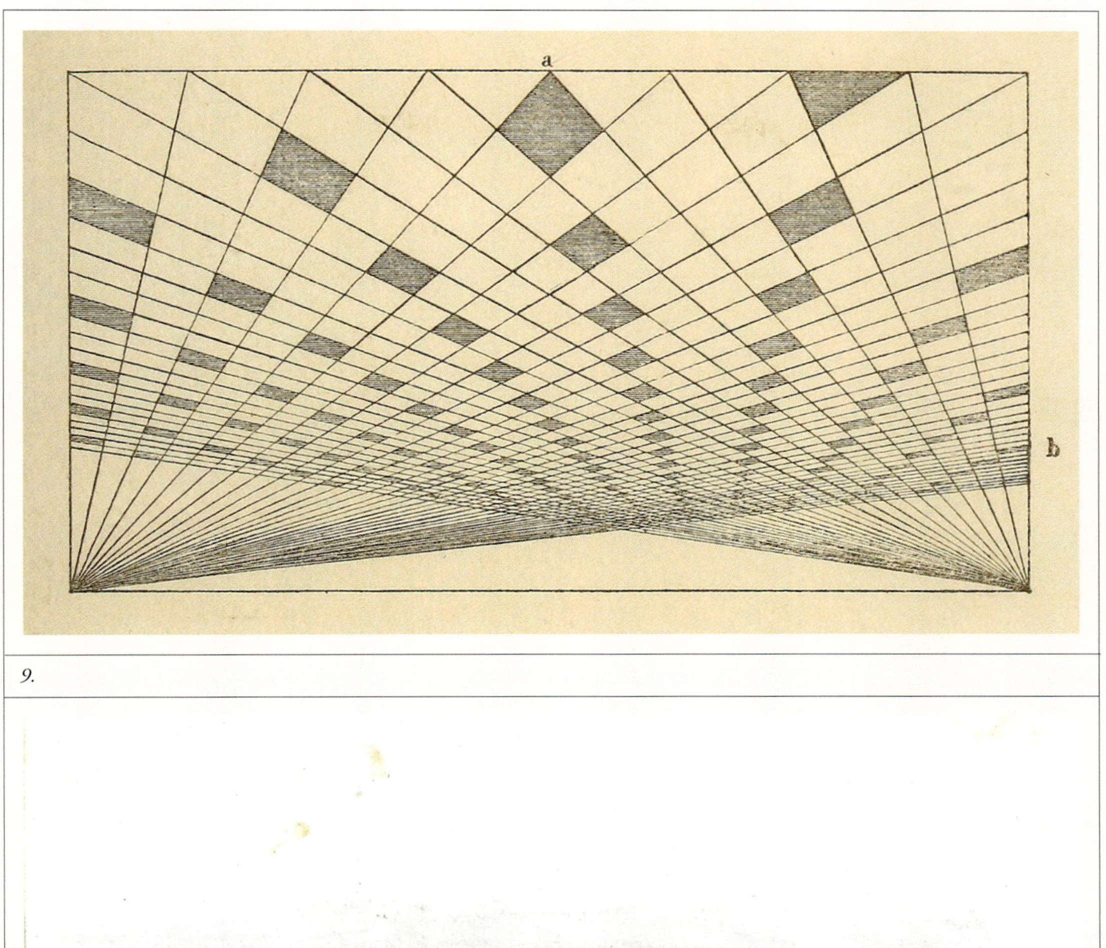

9.

10.

9. **Cloud Perspective (Rectilinear) by John Ruskin, 1860**
In *Modern Painters* (1860), Ruskin proposed a perspectival system for bringing "orderly adherence" to clouds. Most clouds adhere to this geometrical arrangement (though not all).

10. ***Cumulus and Nimbus* by Luke Howard, ca. 1803**
Today, we would classify this as *Cumulonimbus capillatus incus praecipitatio* (a glaciated *Cumulonimbus* with a symmetrical anvil on top and precipitation visibly reaching the ground below).

11.

*11. **Hills and Sky*** **by John Ruskin, undated**

Hill and sky mutually reciprocate each other in this Ruskin classic, depicting conditions close to sunrise or sunset. A bank of low, almost wavy *Stratus* or *Stratocumulus* impinges upon the hill summit (right), perhaps created by the hill itself (species *lenticularis*), due to the air being forced to lift over the hill. Patches of *Stratus fractus* lie on the left. At higher elevation, through the gaps in the lower cloud, we can see a broken deck of (probably) *Altocumulus stratiformis*, with the skylight openings revealing a pale blue half-light beyond.

6

5

4

3

2

1

"It may perhaps be allowable to introduce a Methodical nomenclature, applicable to the various forms of suspended water, or, in other words, to the Modifications of Cloud."

Luke Howard, *On the Modifications of Clouds*, 1803

THE SCIENCE OF CLOUDS

STABILITY AND INSTABILITY

STABILITY

When cold Arctic air moves over warmer ocean waters, moist thermals soon rise, exporting heat and moisture upward to form cloud streets (parallel lines, variety *radiatus*), which gradually grow larger downstream.

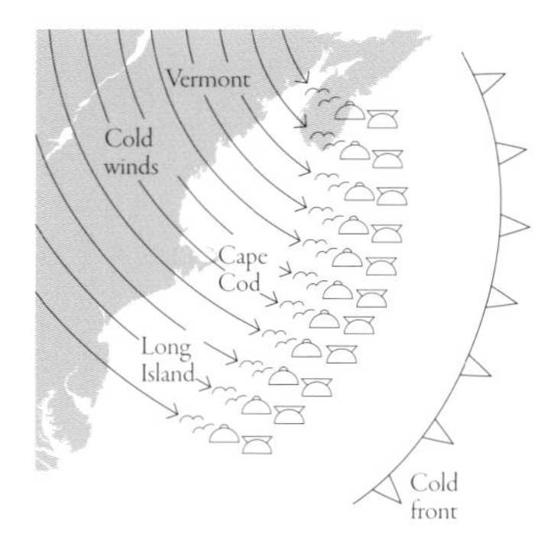

After about 30–50 miles (50–80 km), clouds develop

BUOYANCY

An air-filled ball under water, less dense than water, is forced to the surface.

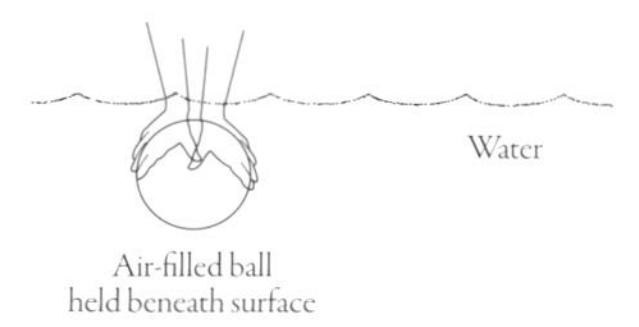

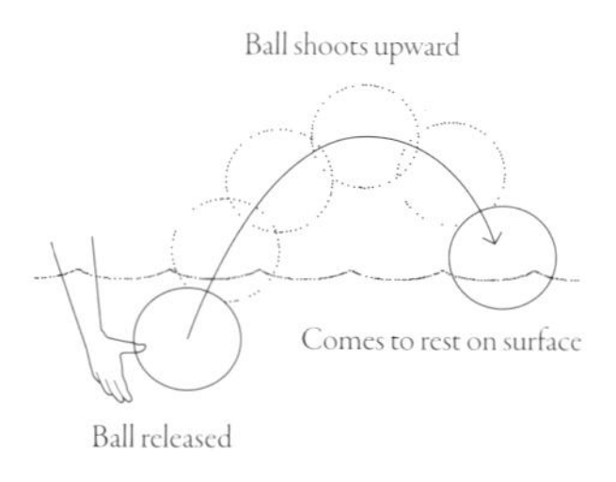

The principle of atmospheric stability is key to the understanding of the formation of clouds, as is its direct counterpart, instability. A stable atmosphere is one in which the normal rate of drop in air temperature with height is reduced, or even inverted, creating a stable layer or an inversion that arrests the vertical development of clouds. In contrast, an unstable atmosphere is one in which the air temperature decreases rapidly with height—if this rate of temperature drop exceeds 5.4°F per 1,000 feet (0.98°C per 100 m) of height, a rate known as the "dry adiabatic lapse rate," dry air will rise of its own volition. We often see this happening on a hot summer's afternoon when warm air currents rise rapidly above a hot area of ground.

The vertical profile of atmospheric stability and instability therefore not only controls whether air starts to rise, but also whether it continues to rise after the formation of a cloud. For example, if air rising in an unstable environment encounters a layer of stable air during ascent, its upward trajectory may be checked, which can cause the rising air to either spread sideways or to sink back down toward Earth and evaporate again.

The principle of stability

In essence, the principle of stability, or rather its direct opposite, instability, is the same as that of buoyancy—warm air will rise of its own volition until it finds itself at the same density as the surrounding air, in the same way that an inflated ball placed underwater finds itself less dense than the water, forcing it to rise to the surface (see diagram, left). Stable air does not rise freely, as it is not buoyant enough. If it is forced upward, for example, by the wind flowing over a mountain, it will tend to sink quickly back downward on the other side, oscillating a little bit in the process.

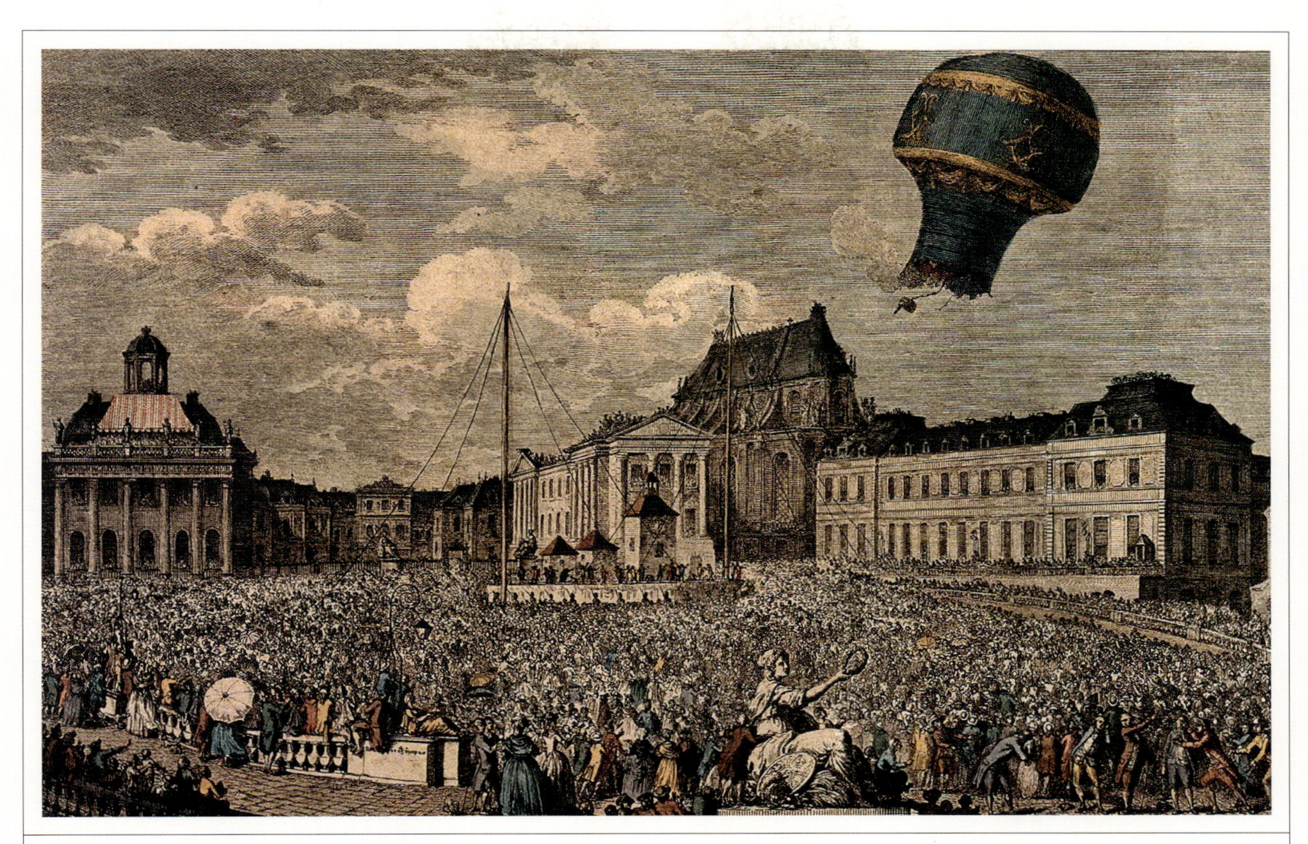

1.

THE DISCOVERY OF ATMOSPHERIC STABILITY AND CLOUDS

By the 1750s, the scientists and natural philosophers of the Enlightenment had discovered the laws of motion (Isaac Newton, 1687), the latent heat of water (Joseph Black, 1750), and that lightning is electrical (Benjamin Franklin, 1751), but they had yet to grasp the full potential of the principle of atmospheric stability. This would take until 1783, when the Montgolfier brothers used a hot-air balloon to exploit the fundamental physics of the atmosphere and demonstrate the world's first aeronautical flight at the Palace of Versailles, in front of Louis XVI and more than 100,000 spectators. A complete lineage of clouds, based on form, shape, height, and texture, would not become established until the next century, after Luke Howard's *Modifications of Clouds* in 1803.

1. **Montgolfier balloon**
Color etching of a balloon ascent by the Montgolfier brothers at Versailles, France, in 1783.

WHAT IS A CLOUD?

CASPAR DAVID FRIEDRICH
(1774–1840)

Along with J.M.W. Turner, Friedrich represents the essence of Romanticism—a movement that rejected the tyranny of reason in favor of the greater wisdom of emotion, of the natural over the artificial. Friedrich deliberately diminished the human presence in his landscapes and insisted on an emotional response to the beauty of the natural world, shorn of the appurtenances of industrial civilization. His most famous painting, *Wanderer Above the Sea of Fog*, presents a merged landscape of sky, clouds, fog, and mountains, all contemplated by an awed yet anonymous human observer.

Have you ever reached out on a mountain top to try to touch a cloud and been disappointed? Or done the same on a foggy day, but feel like you are reaching into nothing? What, then, is a cloud? How can these "airy nothings" appear to form, grow, dissipate, and redevelop right before our eyes? Can they even be defined accurately within a single moment in time? Or, like most things in the natural world, do they represent instead the evolution of a continual process, one which is constantly growing, changing, and then dying, only to be reborn again shortly afterward?

The reality of the matter is that, although we all know what a cloud is, there is no precise scientific definition of a cloud. This is because it is impossible to say—with the exact deterministic precision required of science—when a cluster of cloud droplets or ice crystals has become dense enough to constitute the rather nebulous term "cloud." Were that even achievable, we would still be faced with many other questions, such as how big should that cluster be? How long should it last? Should it be a visible object, or one that is merely perceived? Should it be detectable at other non-visible wavelengths, such as in infrared light? At what intensities and at what limits of these wavelengths? This lack of precision might be viewed as a bit of an embarrassment for science, but perhaps equally as a win for the arts and humanities.

It seems the best definition that can be offered—at least from the scientific perspective—is one that describes clouds as a myriad tiny water droplets or ice crystals, collectively known as "hydrometeors," suspended in the atmosphere and continually in evolution, either visible or perceived, and which act to influence the everyday weather, as we experience it.

2. ***Wanderer Above the Sea of Fog* by Caspar Friedrich, 1818**
This masterpiece from the Romantic movement is largely artistic rather than meteorological, interpreted as a reflection along life's path. Fog "seas," or *nebelmeer* (as they are known in Switzerland and Germany), are commonly found in winter, when the Alps rise majestically above often persistent and stagnant low-level layers of *Stratus*.

WHAT GOES UP...

Allegedly coined in reference to Isaac Newton's law of universal gravitation, we are accustomed to hearing the adage "What goes up must come down," but we do not often think of it in relation to clouds. And yet, how can those immense towering cathedrals of the sky, drifting by silently above our heads, be apparently unaffected by the same laws of gravity that keep both apples and humans tied to Earth's surface?

The scientific reality of the matter may be somewhat surprising—in truth, clouds are always falling down. In fact, they fall down as frequently as they rise up. The trick that nature plays on us, apart from when it rains, is that we do not usually see them "falling down"—usually we see only the opposite, when they rise. This is because clouds become visible when air rises; they are effective tracers of air that either is rising or has risen recently.

Clouds are the manifestation of both the upward movement of air and the coincidental cooling of air. Rising air expands as it encounters lower air pressure on the way up. This expansion causes cooling—something you may notice when you press the valve on a compressed tire— and is called "adiabatic" cooling. The initial warmer air was invisible before it began to rise and cool. However, cool air is unable to hold on to as much moisture as warm air, and the result is that the excess moisture condenses out of the air into tiny droplets or ice crystals, which on a large enough scale are... a cloud.

These nascent cloud droplets are usually very small in size, with typical diameters of only 2–5 microns, or thousandths of a millimeter, which is similar to the sizes of the pollen grains of most common tree species. Like all objects with mass, they experience the pull of gravity. However, given their tiny size, they fall toward Earth at greatly reduced terminal velocities; the largest cloud droplets fall at only a few millimeters per second, a speed that can easily be overcome by the rising air currents inside a developing cloud.

3.

3. ***Cloud Study (Early Evening)* by Simon Denis, ca. 1786–1806**
Cumulus congestus clouds building in an unstable atmospheric environment. The deep blue sky is suggestive of a polar-air outbreak in fall or winter near the ocean. Denis here employed a minimal strip of ground to emphasize the awesome height and scope of a cloud formation. While higher and more distant portions of these clouds are depicted in full sunlight, breaks within them are edged in the pink of early evening. To record these fleeting effects, Denis painted quickly and without making a preliminary drawing. The study was painted in or near Rome.

4.

5.

... MUST COME DOWN

So, clouds are "falling down" all the time, but to us this is mostly invisible. If we do see it happen, it is only when there is a heavy squall or downpour that approaches us rather ominously as a darkened curtain of precipitation, or *virga*. In this situation, the cloud is indeed falling down as rain, hail, or snow, but it is likely building itself up again on the leading edge of the storm not too far away, and may continue to do so repeatedly.

More commonly, clouds simply evaporate, the sinking motion warming the air as they descend due to the greater air pressure encountered, which causes adiabatic warming by compression. In this way, the cloud simply disappears, often doing so quite rapidly. That does not mean the air stops descending at the point where the cloud is no longer visible—we just cannot see its movement anymore.

The descent of air can be initiated by the simple overturning of rising vortices of air thermals, for example in *Cumulus* clouds, which drag down and entrain drier air from above as they rise, allowing evaporation and subsequent cooling, and therefore encouraging a sinking motion. This process leads to the distinctive cauliflower appearance of a well-developed *Cumulus congestus*. Sometimes, the overturning in a cloud can be caused purely by radiational effects—this refers to the process whereby the top of a cloud, usually a layered one such as *Stratus* or *Stratocumulus*, cools directly to space, in the same way as the surface of Earth cools under a clear sky at night. This causes negative buoyancy: the cooled air sinks, bringing drier air downward into the cloud, evaporating it a little.

On a local or more regional scale, air also sinks after crossing a mountain range, which again causes the adiabatic warming of air, and the evaporation of clouds. Windward-facing hills and mountains in the midlatitudes are often much cloudier and wetter than the surrounding lowlands, being very efficient extractors of moisture from the air. The consequence of this, however, is that leeward (downwind) areas are usually much drier and often less cloudy, due to the local descent of air.

LARGE AREAS OF SINKING AIR

Large anticyclones, or areas of high pressure, cover very wide areas—sometimes continental in scale—and are composed of sinking air. In these huge sinking "blocks" of air, the downward velocity of air is very gentle, typically only a few millimeters to a few centimeters per second, but it is usually enough to evaporate most high-level and mid-level clouds, with any low-level clouds kept close to the surface.

The same mechanism operates in the quasi-permanent hemispheric features of the subtropics known as the Hadley Cells, in which upper air that has originated in convective *Cumulonimbus* cells in the tropics moves toward one of the poles and begins to sink at latitudes of around 25–30°N/S.

4. ***At Hailsham, Sussex: A Storm Approaching* by Samuel Palmer, 1821**
Large turrets of *Cumulus congestus* (right) have already built into a heavy shower (center and left), with a pronounced and dark curtain of heavy precipitation (*virga*) advancing from left to right (as indicated by the tilt of the *virga*)—this cloud is certainly falling down dramatically!

5. ***High Clouds Across the Hudson* by Frederic Edwin Church, 1870**
Two *Cumulonimbus calvus* cells tower above the landscape and reflect the evening light. They must be some distance away from the artist as their bases are obscured, although surface visibility is also restricted by a brown haze on the far side of the lake. A few wisps of *Cirrus* or *Cirrostratus* are evident (top right) against the pale blue, humid atmosphere of the evening.

HYDROSTATIC BALANCE

When the atmosphere is in a steady state of balance between the two principal forces acting upon it, namely the vertical pressure gradient force (pushing it upward from high pressure at the surface toward low pressure at altitude) and gravity (pulling it downward, acting toward the center of Earth), we say it is in a state of "hydrostatic balance," meaning it is in vertical equilibrium or "at ease."

Hydrostatic balance is the normal state of affairs in the atmosphere, and it is the reason why horizontal wind velocities on Earth are much greater than vertical wind speeds: usually by two orders of magnitude or more. This situation occurs almost universally everywhere on Earth, the only exception being inside powerful but highly localized thunderstorm updrafts, when vertical wind speeds approach those of storm-force surface wind speeds—but these are rare and brief departures from the norm.

Here is a classic example. When dry, stable air is forced to blow over a range of hills by a steady, moderate, horizontal wind, after ascending to the summit ridge the air will usually descend back to its initial position on the leeward side in a fairly smooth and wave-like fashion. Indeed, the wave motions usually continue downstream in an oscillatory manner for many tens, or even hundreds, of miles, forming "mountain wave" *lenticularis* clouds (page 156) if the tropospheric profile is suitable, before gradually dissipating.

These wave clouds perfectly demonstrate how the atmosphere always tries to return to its normal state of hydrostatic balance, even though it may not happen immediately—just as when a pebble is thrown into a still pond of water, it takes a little time for the oscillation to dampen down before hydrostatic balance is restored.

Schematic showing hydrostatic balance
Air does not "float," and nor is it "as light as a feather." Like all matter, it has a mass, with a density about 1/1000th that of fresh water at sea level. And similar to all objects with mass, the atmosphere is attracted by gravity toward Earth, keeping it close to the surface. At the same time, as air is compressible, atmospheric pressure is greatest at the surface and decreases rapidly with altitude. This vertical pressure gradient tries to push the atmosphere upward, in direct opposition to gravity. The net effect is a quasi-steady state whereby both forces cancel each other out in hydrostatic balance. Small local deviations do occur now and then near strong weather systems and when air crosses hills and mountains, but these are usually dampened out after a few hours—depicted as minor oscillations in the schematic opposite.

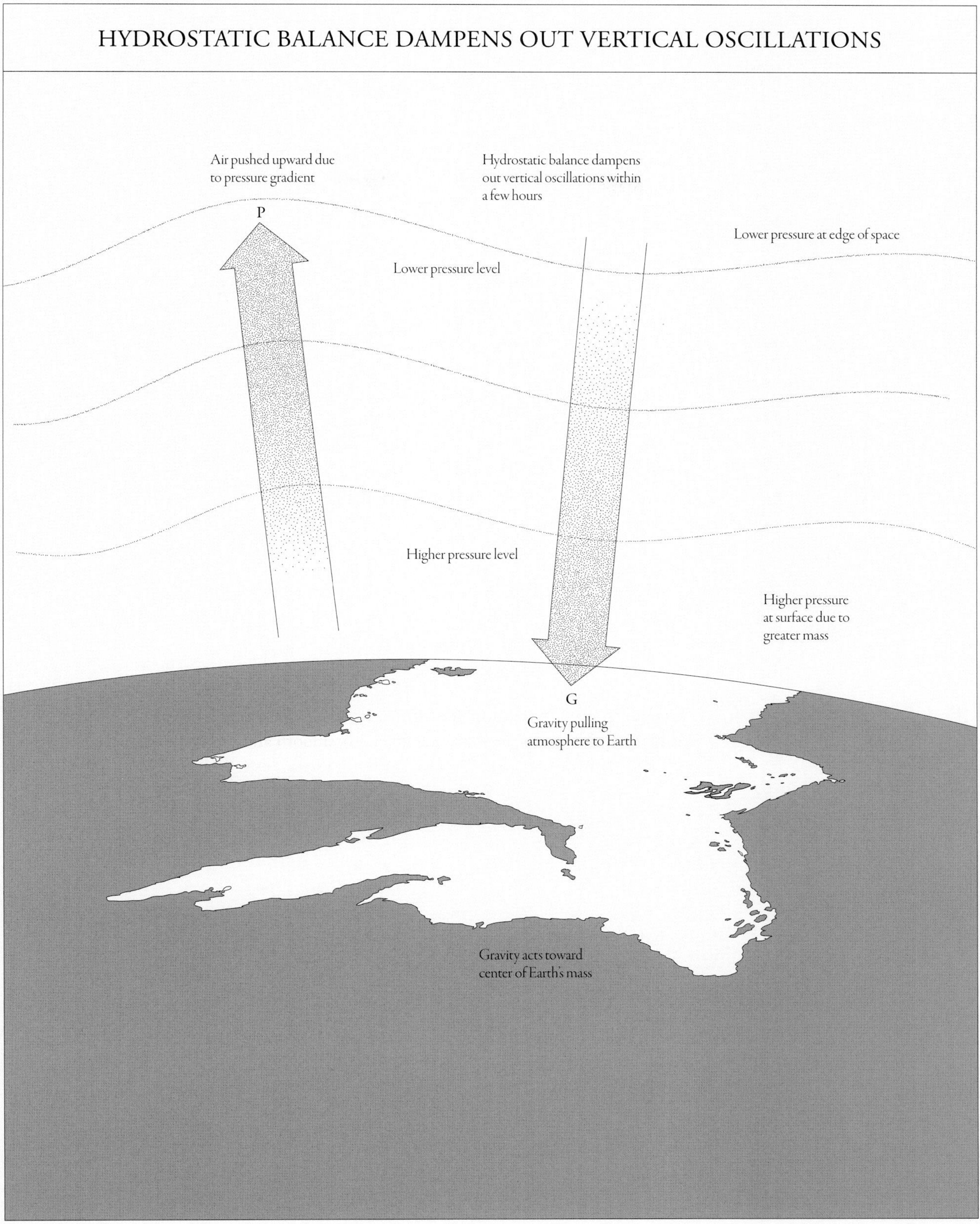

HYDROSTATIC BALANCE DAMPENS OUT VERTICAL OSCILLATIONS

Air pushed upward due to pressure gradient

P

Hydrostatic balance dampens out vertical oscillations within a few hours

Lower pressure at edge of space

Lower pressure level

Higher pressure level

Higher pressure at surface due to greater mass

G

Gravity pulling atmosphere to Earth

Gravity acts toward center of Earth's mass

EVAPORATION OF WATER MOLECULES

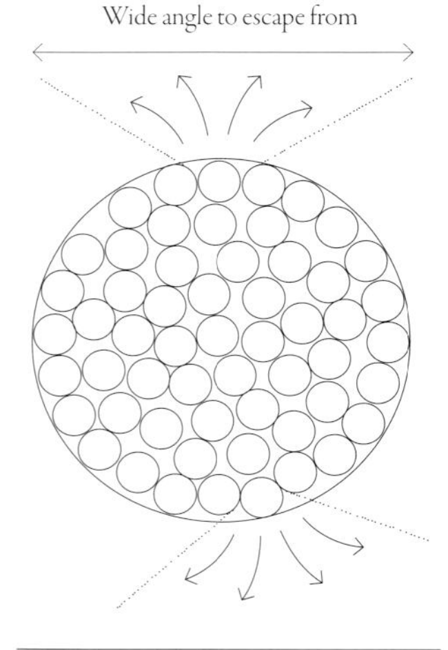

Evaporation
Schematic depiction of water molecules of (Fig. A) water molecules on a planar surface, and (Fig. B) on a curved surface. The molecules in (A) are more tightly bound, whereas in (B) evaporation (escape) is easier. This is why a higher saturation percentage is required to keep droplets from evaporating.

CLOUD CONDENSATION NUCLEI

Strange things happen at small scales. This is true even without downsizing to the quantum dimensions of quarks or the Higg's boson, which has a radius of approximately 10^{-18} m and a mass of about 10^{-27} kg. In cloud formation, it is the size of the smallest constituents involved that is important. These minuscule particles are known as cloud condensation nuclei (CCN).

In their original form, CCN can be solid or liquid, and usually consist of sub-microscopic particles of aerosol such as sea salt, dust, or volatile compounds arising from combustion. Although we cannot see them directly, they are all around us, having an average concentration of around 1 million per gallon (3.8 liters) of air. Without them we would have no clouds nor any rain.

Invisible

CCN originate largely from Earth's surface. They can remain suspended in the atmosphere for many days before falling out or being washed out, and are continuously replenished by land and sea. In terms of mass, most of them are extraordinarily small, weighing around 10^{-16} grams (one-tenth of a quadrillionth of a gram) to 10^{-13} grams (one-tenth of a trillionth of a gram). Size-wise, this means that they are usually shorter than the wavelength of visible light, which is 380–700 nm, or 0.38–0.70 microns. They are therefore invisible to the naked eye and impossible to see using an optical microscope, becoming discernible only under a scanning electron microscope.

What have these tiny particles got to do with cloud formation? Well, it happens that water vapor, as a gas, finds it difficult to condense of its own volition in the free atmosphere, even when the air is heavily supersaturated (having a relative humidity above 100 percent). Indeed, spontaneous condensation of cloud water droplets does not occur until very high—and highly unnatural—supersaturations of several hundred percent are reached.

6.

Escaping from the neighbors

Furthermore, the partial pressure of water vapor, which dictates exactly when condensation or evaporation occurs, is lower over the spherical surface of a curved water droplet than over a flat-water surface (because there are fewer molecular neighbors "holding onto" a molecule on a curved surface than a planar one, and increasingly so for the tiniest droplets—see diagram). This means that any spontaneous condensation of liquid water is more than likely to evaporate again immediately. Water vapor as a gas, therefore, needs a strong helping hand in order to condense into tiny droplets, and for them to persist in the air, so that a nascent cloud may form. These helping hands are CCNs, which act as catalysts in the attraction of water vapor.

6. ***Cloud Study* by Frederic Edwin Church, 1860–70**
A late-evening scene, with a low setting Sun casting its final rays from the right onto the part of a *Cumulus congestus radiatus* cloud street. *Cumulus* are rich in water droplets, absorbing much light—hence the characteristic level bases (page 98) of the cloud appear dark and threatening. The airmass is clean and unpolluted, evidenced by the polar blue sky background.

THE EQUILIBRIUM RADIUS

CCN are crucial in the creation of embryonic cloud droplets. Acting similarly to chemical catalysts, CCN greatly speed up the process of condensation from gas to liquid by being "hygroscopic" (having the ability to absorb water vapor from the air). Water vapor is therefore attracted to the aerosol, even at relative humidities well below 100 percent saturation.

For example, in coastal areas where the humidity is often high, and where this is a steady supply of oceanic aerosols, such as sea salt, water vapor will condense on such CCN at relative humidities of approximately 78 percent and above, initially forming a haze or mist, which restricts visibility. Over land and continental areas, the same process will also occur on tiny dust or clay particles. This permits the aerosols to grow significantly in size as they scavenge water vapor from the air.

Tipping point

If the relative humidity continues to increase to saturation, or even a slight supersaturation—a little over 100 percent—the CCN will continue to grow until an "equilibrium radius" is reached (see illustration). Such CCN are now said to be "activated" and a stable cloud droplet is born. In equilibrium, typical radii range from a few tenths of microns to a few microns, depending on the type and mass of the original aerosol—they can be as large as 20–30 microns for "giant" aerosols.

When this tipping point is reached, the level of supersaturation required to keep the droplet growing actually decreases as the droplet gets bigger (see illustration, where curve starts to decrease to right). This means that droplet growth is now unimpeded, and it can keep on growing as long as adequate moisture is available. In practice, though, this is not always the case, as other droplets nearby will be competing for the same moisture—as such, if a cloud is to be sustained and generate rainfall, a supply of water vapor needs to be maintained.

Droplets with ambition

So if our growing droplet is to have any ambition of becoming part of a rain cloud, a tentative balance needs to be struck between the rate of production of available water vapor—maintained by the lifting and adiabatic cooling of the cloud itself—and the removal of water vapor through condensation, as well any mixing caused by entrainment of dry air from outside the cloud. If these processes are sustained, it will take at least 20 minutes or so for the cloud droplets to grow by diffusion to a size of 10 microns in radius, even in the most rapidly developing convective clouds (*Cumulus* or *Cumulonimbus*; pages 90–101 and 126–135, respectively).

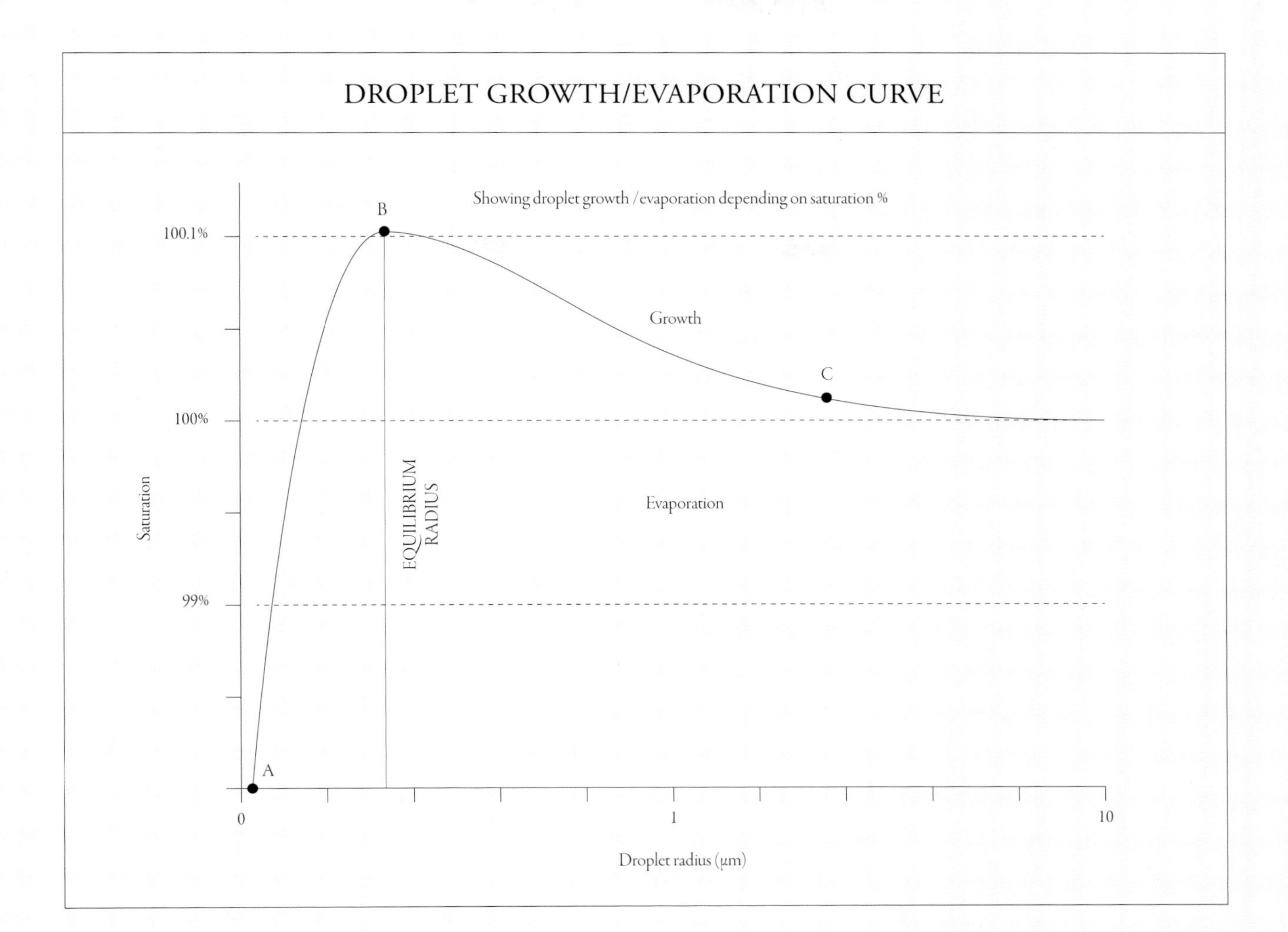

Humidity
As the relative humidity increases from below 99% (point A) to a slight supersaturation of 100.1% (point B), the cloud droplet will grow in size. If, however, the air humidity falls before the droplet reaches point B in size (its "equilibrium radius"), the droplet will start to evaporate. Once the droplet surpasses its equilibrium radius (point B), it can continue to grow, even at lower humidities than at B, for example as indicated by point C on the growth curve. If the humidity drops to values below the curve, however, the droplet will start to evaporate again.

A

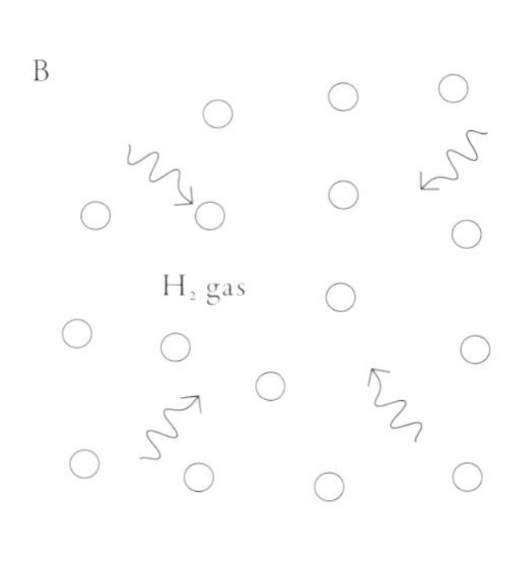

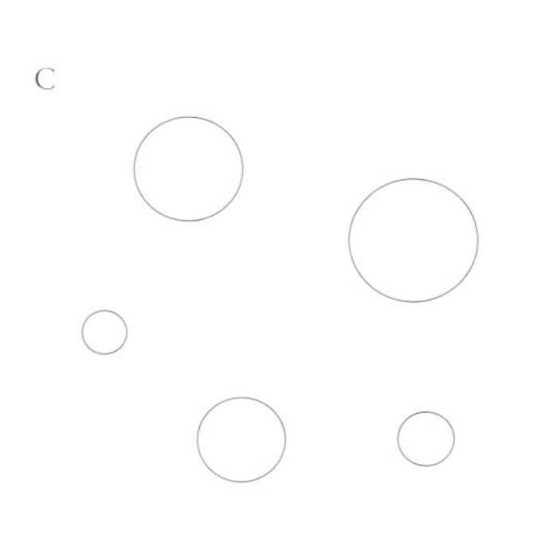

Cloud condensation
Schematic depiction of:
(A) Cloud condensation nuclei (CCN); (B) Water vapor being attracted to the hygroscopic CCN by diffusion and condensing on their surfaces; (C) Activated large cloud droplets after growth by diffusion and collision and coalescence.

DROPLET GROWTH BY DIFFUSION

The process that controls how nascent cloud droplets start to grow on CCN is the diffusion of water vapor. Whether we know it or not, we are all familiar with the principle of diffusion: for example, when we enter the door of a café, bakery, or pizzeria, we are usually greeted by a pleasant warm aroma, which is caused by the diffusion (and also convection) of fragrant gases emanating from a kitchen oven or coffee machine, moving toward our noses. In this case, diffusion can be explained as the random movement of gas molecules, which naturally mix and spread out from regions of high concentration to low concentration over time.

In the same way, where there is a high concentration of water vapor—high relative humidity—in the air, gas molecules diffuse from it onto tiny CCN, the process being initiated first by the water-attracting hygroscopic aerosols. The aerosols then grow rapidly in size (within microseconds), as long as water vapor is available. However, as other aerosols nearby will be competing for the available water vapor, such growth is not assured.

Although extremely fast to begin with, the rate of diffusion of water vapor onto a cloud droplet starts to decrease rapidly as the droplet radius increases beyond 2–3 microns—it slows down considerably and takes many minutes more to grow larger than 5 microns, and many hours to grow beyond 20 microns. This is because diffusion becomes increasingly inefficient as the droplet's surface area increases (the surface area of a sphere [$4\pi r^2$] increases in proportion to the square of its radius [r]). The direct consequence of this slowdown in the growth rate of cloud droplets is that most clouds never rain.

Clearly, other processes acting much faster than a few hours are required if a droplet is to continue toward its ultimate destiny of becoming a raindrop. But only a few cloud droplets will ever achieve this, and they are associated with a few specific, special cloud types, usually *Nimbostratus* (page 122), deep *Cumulus* (pages 90–101), or *Cumulonimbus* (pages 126–135). In most non-precipitating clouds, the cloud droplet radius tends to remain in the 5–10 micron range, with a global average of around 6 microns. Such clouds simply do not have sufficient numbers of large droplets to create precipitation—or even if they do briefly, the clouds are too thin and too tenuous for them to endure, or their cloud lifetimes are too short, and any aspiring, suitably sized droplets evaporate soon after leaving the cloud.

A GROWING DROPLET

There is yet more to come in the existential voyage of a nascent cloud droplet as it attempts to attain its ultimate destiny—one necessary for human life on Earth—that of a raindrop falling to Earth. First, the droplet must grow to a radius of approximately 10–20 microns if there is to be any hope of rain developing, and it needs to do so by diffusion alone. If we pick one droplet at random, such growth is unlikely to occur by diffusion alone, as it would take several hours to achieve a radius of this size. However, if we take a statistical approach and consider the cloud as a whole, which is composed of quadrillions to quintillions of cloud droplets, the chances are much greater that at least one droplet out of all of these will have achieved the necessary size, due to random collisions between the cloud droplets. Some droplets simply get lucky.

Pinball and dodgems

It is also at this crucial stage of cloud and droplet growth, that other important effects begin to take over and make their influence felt. The principal one is a force that we are all familiar with: gravity. Gravity starts to become significant now because as a cloud droplet's radius surpasses 10 microns, the effects of air resistance start to diminish appreciably, with the cloud droplet's fall velocity increasing to more than 1 cm (0.4 in) per second. This means that it now has the potential to fall through the cloud. And in an ascending updraft of air, it will ascend at a slower rate than its smaller neighbors, increasing the chance of collision. Once a certain proportion of drops reach beyond this critical radius, a game of celestial pinball begins, with the largest cloud drops falling faster (or ascending more slowly) than the smaller ones, causing them to bump into one another like dodgem cars, colliding and coalescing, or breaking up (unlike dodgem cars, thankfully), in a sort of unmitigated chain reaction. As the large cloud drops collect the smaller droplets, they continue to become larger still, and fall even faster, thus increasing the chance of ever more collisions.

It turns out that cloud droplets in oceanic clouds, which contain "giant" sea-salt CCN, are more likely to reach a critical radius of 10–20 microns within a time frame that is considerably shorter than the lifetime of the cloud itself. This has important consequences for the development of rainfall and explains why *Cumulus mediocris* (page 94) and some *Stratocumulus* (page 112) clouds are more likely to produce precipitation over oceanic and coastal environments than over continental areas. Precipitation can also develop quickly within towering *Cumulus congestus* clouds after approximately 20 minutes, as soon as suitably sized droplets are produced by the diffusion process. Then, the "collision and coalescence" mechanism quickly takes over.

7. ***Clouds* by Thomas Cole, ca. 1838**
A powerful *Cumulus congestus*, likely soon to grow into a *Cumulonimbus calvus* (a non-glaciated *Cumulonimbus*). The characteristic bright white and fractal cauliflower appearance of the rising cloud surface is caused by the cloud's high water content, together with the entrainment of drier air from above as the cloud tower rises. On close examination, there appear to be two or three rings of cloud encircling the cloud summits—these are the variety *velum*, found only with rapidly ascending *Cumulus* or *Cumulonimbus*.

COLLISION EFFICIENCY

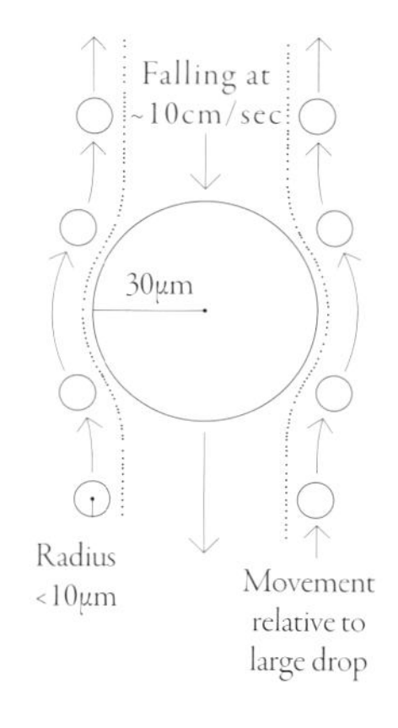

Large cloud droplets fall much faster relative to very small ones, sweeping them aside as they fall.

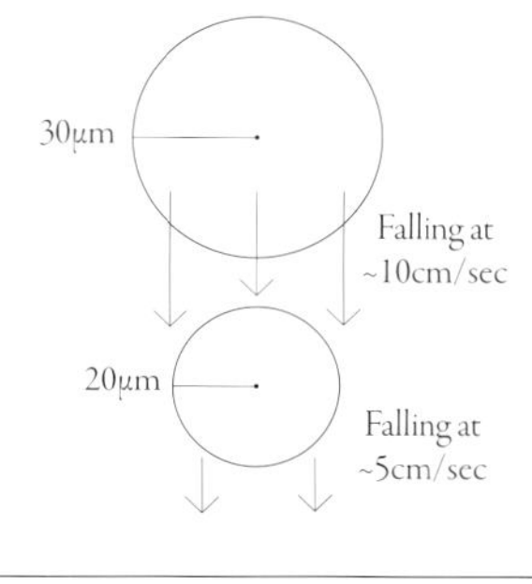

Similarly sized large droplets are more likely to coalesce together. Here, the faster-falling larger droplet collects up and joins together with the slightly smaller droplet.

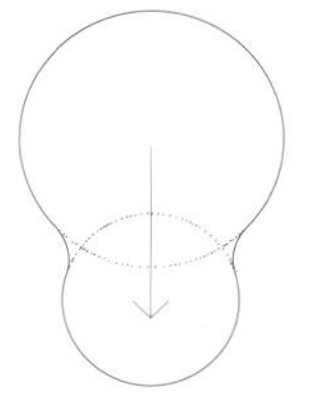

The two droplets combine to form a single larger droplet. If conditions remain suitable in the cloud and there are many more collisions, eventually this droplet will fall from the cloud as a raindrop.

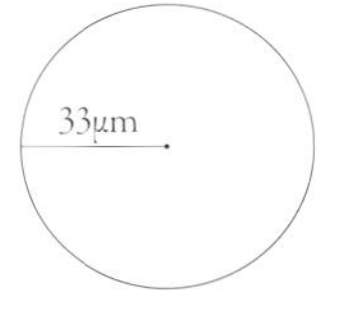

BIG DROPS FALL FASTER

It transpires that the collision efficiency of cloud droplets—that is, whether they join together after colliding with one another—depends strongly on their absolute size. The largest drops have the highest collection efficiencies, meaning that they are more likely to coalesce with each other. This is because air has a certain degree of "viscosity," or cohesiveness, which allows small droplets to be swept aside by the air currents surrounding them, but this process becomes less effective for larger droplets. In fact, cloud droplets with radii below 10 microns rarely coalesce. Not all large droplets coalesce, however: they may also split each other apart when they collide.

In the consequent melee of a rapidly developing cloud in which droplet collision and coalescence has begun, we now need to start considering the different fall velocities of the various drops and droplets, both relative to one another and also relative to ascending and descending air currents within the cloud. These are summarized in the table opposite: as one might expect, the larger the droplet, the faster it falls.

When cloud droplets grow beyond a radius of approximately 30 microns (0.03 mm), they are referred to as "drizzle drops." This size coincides with fall speeds of greater than about 4 inches (10 cm) per second: greater than the speed of ascending air currents in most layer clouds, such as *Stratus*. This partly explains why we often experience drizzle from low-level clouds on a damp, humid day, but not from higher-level layer clouds. For example, a drizzle drop of 100 microns (0.1 mm) radius that exits a cloud base at an altitude of 650 feet (200 m) and falls at a rate of 1.8 miles per hour (2.9 km/h) will reach the ground within 4 minutes (see table). However, any drizzle from mid-level cloud is much more likely to evaporate during the longer descent time.

At the other extreme, the largest raindrops fall at speeds of up to 25 miles per hour (40 km/h). At these velocities, they are no longer spherical because they are deformed by the air as they fall, becoming oblate in shape. Raindrops with a radius larger than 3–4 mm are unstable and tend to break apart as they fall.

8. ***Dark Cloud Study* by John Constable, 1821**
This cloud may be dark as seen from below, but only because it is rich in water droplets. Its upper surface is likely to be very bright, reflecting and scattering most of the Sun's rays. The cloud is convective—the texture suggests building heaps of *Cumulus congestus*.

8.

Terminal velocities of cloud droplets, drizzle, and raindrops

RADIUS (MICRONS*)	TERMINAL VELOCITY	NAME	TIME TO FALL 1 KM (0.621 MILES)
0.1	0.001 mm/sec	Cloud condensation nuclei	----
1	0.12 mm/sec	Cloud droplet	----
5	2.9 mm/sec	Cloud droplet	----
10	1.2 cm/sec	Cloud droplet	23 hours
30	10.7 cm/sec	Drizzle	2 hours 35 minutes
100 (0.1 mm)	80 cm/sec	Drizzle	21 minutes
300 (0.3 mm)	2.4 m/sec	Raindrop	7 minutes
3000 (3 mm)	10 m/sec	Largest raindrops	90 seconds

*One micron is one-thousandth of a millimeter (0.00004 inches)

THE DROPLET FAMILY

Typical CCN
n=300,000
r=0.1
v=0.001

Typical cloud droplet
r=6
n=100,000
v=0.4

Large cloud droplet
r=50
n=1000
v=30

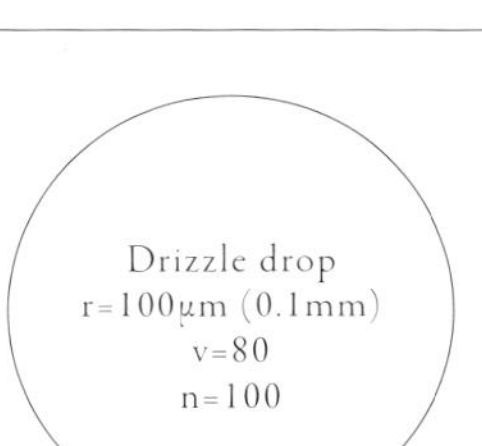

Typical raindrop
r=100µm (1mm)
n=1
v=650

r radius (microns)
n number per liter
v terminal velocity (cm/s^{-1})

THE CLOUD DROPLET SPECTRUM

The total number of collisions that a single drop can make during its journey toward Earth can be upward of tens of thousands. This statistic is more easily appreciated looking at the illustration (left), which shows the comparative sizes of a cloud condensation nucleus, a freshly activated cloud droplet, a drizzle drop, and a typical raindrop. Bearing in mind that the volume of a sphere increases with the cube of its radius—the volume of sphere (V) with respect to its radius (r) is given by $V=\frac{4}{3}\pi r^3$—the volume of a 1 mm radius raindrop is therefore 1 million times greater than that of an activated cloud droplet of 10 microns in radius.

Newly formed clouds tend to have a much greater concentration of cloud droplets than older clouds. This is because the droplets have not yet had sufficient time to grow into larger droplets: a process that reduces the total number of droplets but increases their average size and volume. Hence clouds such as *lenticularis* tend to have high numbers of very small droplets (page 156). On the other hand, relatively "old" clouds, such as *Stratus* or marine *Stratocumulus*, tend to have lower droplet concentrations, but with a larger average size.

As already mentioned, due to differences between the type of CCN available in oceanic and continental regions on Earth, maritime and coastal clouds tend to have lower cloud droplet concentrations, but with a larger average droplet size than their continental equivalents. In maritime areas, this is thought to be due to the presence of giant aerosols such as sea salt, which scavenge the cloud environment of its tiniest droplets. Having larger activated cloud droplets therefore increases the chance of precipitation, as collision and coalescence can start earlier than in other clouds.

Here comes the rain

Returning to our cloud droplet: once a critical radius of about 10–20 microns is reached, the dual processes of collision and coalescence now start to take over. A rapid broadening of the droplet spectrum ensues, as the larger drops quickly sweep up the smaller ones. If the cloud is shallow in depth, such as in a relatively thick layer of *Stratus* or *Stratocumulus* close to the surface, a slight drizzle may start to fall. Alternatively, in the case of rapidly developing convective clouds such as large *Cumulus* and *Cumulonimbus*, a few fat drops of rain may start to splash down, as these are the only drops large enough to escape the rising updrafts of the ascending cloud.

Recent research increasingly points toward the entrainment of dry air from immediately outside the cloud environment, drawn into the cloud and diluting it as it ascends, as being instrumental in the development of a broad or multi-modal droplet spectra in a convective cloud—a characteristic of mature rain-bearing *Cumulonimbus*. This happens because when dry air is brought into the cloud, the smallest droplets will evaporate first, leaving behind only the larger ones; these may, after a brief period of descent, start to ascend again in a fresh updraft a few minutes later, when they will begin to grow again to an even larger size. This process may be repeated many times over in a large *Cumulus congestus* or in a *Cumulonimbus* cloud, sorting the raindrops (or hailstones) quite efficiently in a mechanism described by the English cloud physicist Sir John Mason as a giant "winnowing" machine (in reference to machines that separate the lighter chaff, or seed casings, from heavier plant or cereal grains). The net result of this is a rapid broadening of the droplet size distribution or spectrum with a long tail toward the right (see illustration below) with occasional multi-modal peaks.

However, these processes are reserved for only a few clouds, those producing precipitation—the fact of the matter is simple: most clouds never rain. Should we spare a thought for the countless quintillions of cloud droplets that "never make it"? All those tiny droplets that never get near the raindrop stage? No need! These so-called "non-achieving" droplets make up the vast majority of clouds that we see and marvel at. Together, they create nature's celestial landscape right before our eyes. Each droplet, big and small, momentary or otherwise, has its own place in the magnificent orchestral tableau of the skies that we call "clouds."

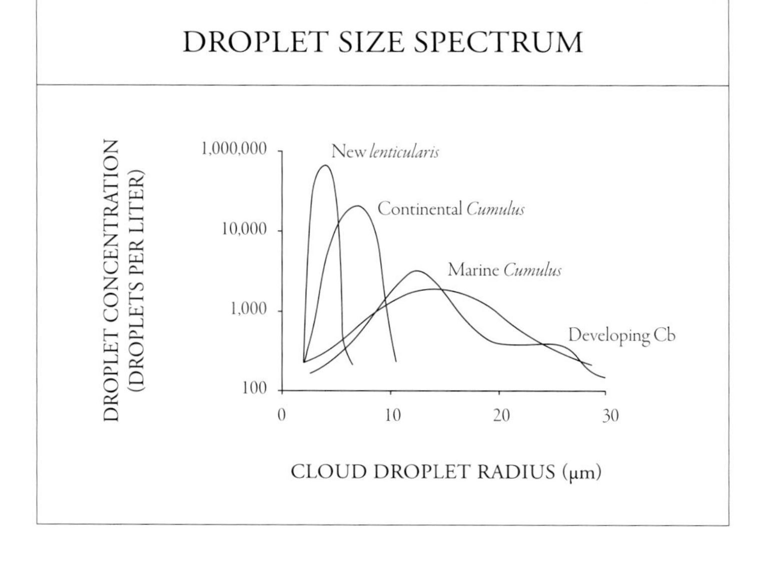

Opposite: Schematic illustration of comparative sizes (scaled relatively) of a typical cloud condensation nuclei, cloud droplet, large cloud droplet, drizzle drop, and raindrop. Also listed are their typical radius (r), concentration per liter (n) and terminal velocity/fall speed (v). Values are based on Macdonald (1958) and Mason (1975).

Left: Schematic example of the droplet size spectrum (their statistical distribution) for four different clouds: *lenticularis* (freshly formed); continental *Cumulus*; marine or oceanic *Cumulus*, and a developing *Cumulonimbus*. The plotted values are approximate: cloud droplet spectra are continuously changing within clouds; no two clouds have the same spectrum.

9. ***Staffa, Fingal's Cave***
by J.M.W. Turner, 1832
It's a common enough squally day on the west coast of Scotland. A shaft of heavy precipitation (right) falls from a passing *Cumulonimbus*, the tall cloud creating great contrasts between light and shade. Surface visibility appears restricted to only a few miles by buoyant sea spray, which lends a pale yellow hue to the watery skies. The sea is raging, but the steamer boat appears to be coping; its lofting, dispersive plume tells us both the direction of the gale and alludes to the unstable but well-mixed state of the lower atmosphere.

GLACIATION

Stage 1 Water droplets only

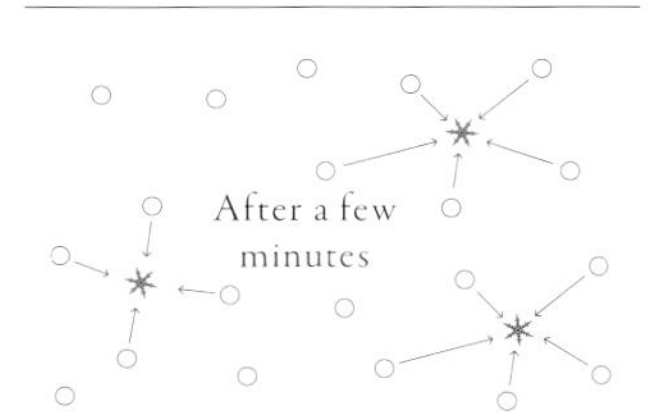

Stage 2 Ice crystals develop

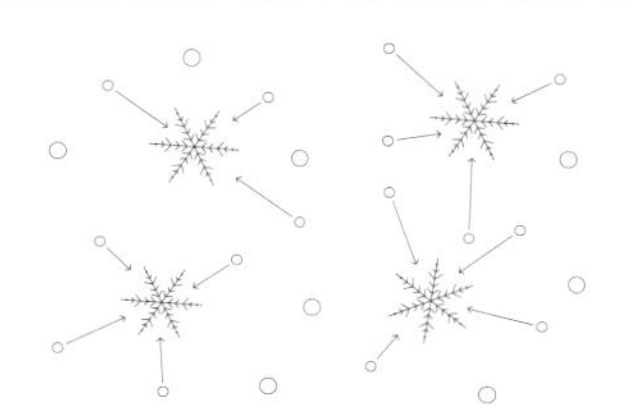

Stage 3 Ice crystals grow at expense of water droplets

Stage 4 Glaciation complete

Glaciation stages
Stage 1: Supercooled water droplets suspended in a cloud with a temperature of well below freezing and at a humidity saturated with respect to ice. Stage 2: Introduction of ice nuclei, allowing the commencement of freezing and growth of ice crystals at ice nucleation sites. Stage 3: Rapid growth of ice crystals at the expense of cloud droplets. Stage 4: Glaciation complete, the cloud now consists of ice crystals only, which, if large enough, will begin to precipitate toward Earth.

ICE CRYSTAL GROWTH

So far we have only dealt with *liquid droplets* that lead to the formation of rain within clouds in a mechanism described by atmospheric scientists as the "warm rain" process, even if the droplets involved are really quite cool or indeed cold (but not frozen). This warm rain process is common in tropical and subtropical climates, and in warm seasons elsewhere, especially at low and midlevels in the troposphere. However, away from the tropics, a far more important rain-producing mechanism is often at work, mainly at mid- and upper levels of the troposphere, and particularly during the passage of weather fronts and cyclonic weather systems whose coherent cloud structures often reach up to these levels.

Contrary to intuition, supercooled liquid water droplets can exist quite freely in clouds right down to temperatures of -4°F (-20°C) and below, as spontaneous freezing of all liquid water droplets in a cloud does not occur until an ambient temperature of -36°F (-38°C) is reached. However, if a frozen ice crystal or a "seeding particle" is introduced into a strongly supercooled cloud, freezing will usually take place immediately, and spread throughout the cloud within a few minutes in a rapid chain-like-reaction (see *cavum*, page 199). This is because the degree of saturation required for ice crystals to form in a cloud is lower than its equivalent for liquid water droplets (see illustration, left)—at an air temperature of -22°F (-30°C), when the relative humidity with respect to ice is 100 percent (saturated), it is only 75 percent with respect to water (unsaturated), so only ice crystals can form. This important attribute arises because of the stronger intermolecular bonds in ice crystals, compared to those in liquid water droplets.

Once ice crystals start to grow in very cold clouds, they do so through the diffusion of water vapor onto their crystal surfaces *at the expense of* liquid water droplets. The ice crystals therefore grow quickly, scavenging nearby water droplets as well as additional moisture from surrounding air, because the humidity threshold required for their growth is much lower than for water droplets. Soon, these crystals will be large enough to fall to Earth, growing even bigger as they pass through other clouds on their descent (again, by scavenging their water), before finally melting into raindrops in the final few thousand feet, if the surface air temperature is a few degrees above freezing point.

10.

A large range of ice crystal shapes, or "habits," exist, such as hexagonal plates, ice columns, needles, and small dendritic snowflakes. Large snowflakes do not aggregate together until the lowest levels of the troposphere, close to the melting point of 32°F (0°C). This is why snowflakes tend to remain small during the coldest snowstorms, but are large and feathery in heavier and wetter falls, which usually take place at an air temperature near freezing point.

The cold precipitation mechanism described here is known in meteorological circles as the Wegener–Bergeron–Findeisen process, after the three scientists who first theorized it in the early twentieth century, namely Alfred Wegener, Tor Bergeron, and Walter Findeisen. It is a critical and fundamental part of the meteorological operation of the atmosphere, without which we would, quite literally, die of thirst, because much of the precipitation falling across the middle latitudes originates in this way.

10. ***Ice Clouds over Coniston Old Man*** **by John Ruskin, 1880**
A gap in the cloud layers reveals high-frequency *undulatus* billows (top left), as well as dark lower-frequency *Stratocumulus lenticularis* mountain waves (lower third). These mimic the general topography but appear to obscure most of the mountain view. Some of the other waves (left of center) appear to be an expressionist depiction of local turbulence. The low-level *Stratocumulus* clouds are certainly not frozen, although the higher white clouds (upper center) may represent icy *Cirrocumulus* or *Cirrostratus*.

CLOUDS AND RADIATION

Absolute zero—zero Kelvin, or -459.67°F (-273.15°C)—is the lowest possible theoretical temperature. All objects that have a temperature above it therefore contain some energy; as a result, they will emit some form of electromagnetic radiation, such as microwaves, visible light, or X-rays. The type of radiation an object emits depends on how hot it is, according to a relationship known as the Stefan–Boltzmann law.

As the Sun's surface temperature is approximately 10,800°F (6,000°C), the Stefan–Boltzmann law states that solar radiation emitted from it peaks at visible wavelengths of about 500 nanometers (0.5 microns, near what we call the color blue) but also covers a large proportion of the ultraviolet and near infrared parts of the electromagnetic spectrum. In contrast, radiation from objects at everyday temperatures—including terrestrial radiation from Earth and its envelope of clouds, which largely lie in the range of approximately -100°F to +100°F (-73°C to +38°C)—is almost entirely infrared, peaking at about 10,000 nm (10 microns, or 0.01 mm). Because of its longer wavelength, and therefore lower frequency, infrared radiation carries less energy than solar radiation.

Clouds, like any other object, intercept, reflect, absorb, and re-emit different forms of electromagnetic radiation. This can happen in a variety of ways. Regarding solar radiation, clouds may intercept a proportion of it and absorb it, or, depending on their albedo (reflectivity), they may reflect it. (Some upper cloud surfaces are brighter than others—page 73—which also means some cloud bases are darker than others too). As all clouds have temperatures within the modest terrestrial range as stated above—they principally emit infrared radiation both to the ground and the sky above, or to other clouds nearby. These clouds will then absorb and re-emit infrared radiation themselves in the same manner.

Humans are able to perceive only a very small fraction of the electromagnetic spectrum, namely visible light (300–780 nm, or 0.3–0.78 microns), which we can see, and infrared radiation, which we feel as heat gained or heat lost. Without the necessary equipment and instrumentation, it is impossible for us to detect most other types of electromagnetic radiation, although we may experience its consequences, for example when ultraviolet radiation causes sunburn.

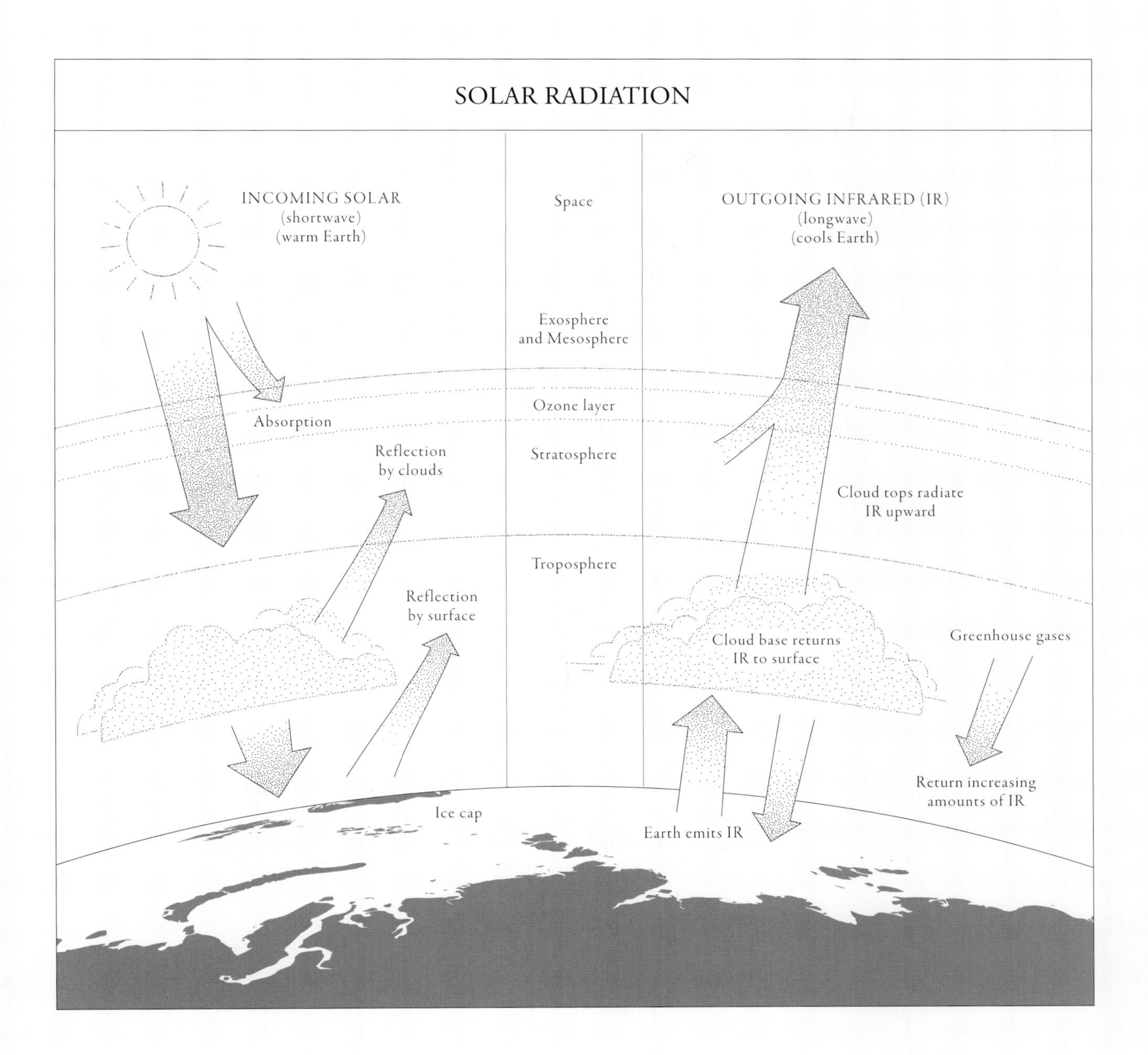

Solar radiation
Schematic depicting solar radiation exchanges in the atmosphere (left) and infrared radiation exchanges (right).

If you sit very still and watch closely, over a period of an hour or more, you can sometimes observe the effect of infrared radiation from the upper surfaces of layer clouds. For example, a featureless, uniform layer cloud such as *Stratus* or *Altostratus* will gradually morph over time into a slightly dappled layer of *Stratocumulus* or *Altocumulus*. This occurs as a result of two radiation processes: cloud top cooling to the clear sky above (assuming no higher cloud layer prevents this from happening), and cloud base warming from Earth's infrared radiation below. This creates gentle convection currents that help to overturn the cloud, creating a dappled effect of regular cloudlets.

JOHN CONSTABLE
(1776–1837)

The distinctive landscape of southeastern England has been known as "Constable country" for more than two centuries. Yet many scholars have argued that Constable's obsessive fascination with cloud formations charts the baleful effects of climate change. Unlike the more expressive cloud paintings of some of his Romantic contemporaries, Constable made around one hundred meteorologically accurate cloud studies between 1821 and 1822, giving modern environmental scientists a clue to the changes we witness in the modern landscape.

HAZE, SMOG, AND FOG

Haze is a broad meteorological term that describes reduced visibility. It is typically used by meteorologists to describe tiny solid particles suspended in the atmosphere, such as aerosols consisting of dust, sand, volcanic ash, soot, smoke, or other byproducts of combustion. Its brown or orange hue is commonly seen over cities during periods of settled weather.

In a more general sense, haze also describes reductions in visibility caused by tiny liquid particles, such as sea spray, or activated hygroscopic cloud condensation nuclei, although when the horizontal visibility falls below 5 miles (8 km), the term "mist" is preferred by meteorologists.

The definition of "fog" is much stricter, referring to a horizontal visibility of less than exactly 1 km (0.621 of a mile) at ground level. Fog usually manifests itself as a result of local factors (for example, when moist air pools in low-lying areas during still weather, thickening later into a layer of *Stratus*). Alternatively, in coastal areas, it can be advected (transferred) onshore by the breeze, usually as *Stratus* or *Nimbostratus*.

"Smog" is a colloquial term describing an atmospheric soup of "smoke" and "fog" mixed together; strangely, despite its frequent occurrence in (industrial) Victorian Britain, the term only began to be used widely after 1950. Smog tends to be a winter phenomenon. On the other hand, "photochemical smog"—a term reserved for a particular noxious chemical cocktail of air pollution and haze—forms over cities in bright summer sunshine and is highly deleterious to health. Both smog and/or photochemical smog are nearly ubiquitous in large cities such as Los Angeles, Mexico City, Beijing, and Delhi, and many others, depending on the time of year and season.

11.

The effects of air pollution

There are a multitude of issues that arise from air pollution—for example, according to the World Health Organization in 2023, 6.7 million premature deaths annually across the globe are caused by the combined effects of outdoor (ambient) and indoor (household) air pollution. Beyond this, the large increase in local cloud condensation nuclei (CCN) associated with air pollution can also have an indirect effect on the climate. A cloud may become brighter and reflect more sunshine (due to a higher reflectivity, or albedo, appearing whiter to the eye when seen from above) if the total number of nuclei is suddenly increased. This cloud-brightening process is known as the "Twomey effect," after the Irish American scientist Sean Twomey, who first described the mechanism in 1967; it has gained renewed interest in recent years, thanks to various climate intervention and geoengineering proposals to slow down or halt global warming by making clouds brighter (more reflective of sunshine).

In highly polluted environments, the concentration of aerosols acting as CCN may be so high—of the order of tens of millions per gallon (or tens of thousands of particles per cubic centimeter)—that they actively "scavenge" all of the available water vapor from the air, and in doing so prevent any CCN from being activated (reaching its equilibrium radius). The net effect is to suppress cloud formation, although any benefit from reduced cloudiness is likely compromised by a severe reduction in air quality.

11. ***Hampstead Heath Looking Toward Harrow*** **(cropped) by John Constable, 1821–22**
At the time of this painting, London was a much smaller city than it is today; even so, the air looks hazy, dusty, and polluted. The westerly evening panorama is suggestive of a few smoky plumes originating from ground level in the distance, with a likely abundance of large aerosols contributing to the orange and brown hue of the atmosphere. (See page 184 for full painting.)

LIGHT AND SHADE

When light from the Sun enters Earth's atmosphere, it becomes subject to a range of physical processes and optical effects. Initially, some of the light may be reflected directly back out to space by smooth or bright surfaces that have a high albedo, such as from the tops of bright clouds, or when the ground is covered with snow and ice. In these cases, only a minority of the light is transmitted through, or into, the substance in question; most of it is reflected back directly into space. In contrast, darker surfaces, such as forests, woodlands, or a rough sea, absorb much greater amounts of incoming light, reflecting only a little—this is why we perceive them as dark.

In addition, the texture of the reflecting surface controls how incoming light is reflected in different directions. "Specular reflection" is mirror-like, and therefore unidirectional, such as when we see a perfect reflection from a smooth water surface or pond. However, clouds and most objects on Earth are irregularly shaped, causing light to be reflected diffusely in multiple directions—this light is said to be "scattered." Most objects that we can see and distinguish, as we go about our everyday lives, are visible because light is scattered from them.

Scatterers of light

We see clouds because they scatter light toward us. The ones that are brightest when viewed from above tend to be among the darkest when viewed from below. This is because by the time the light reaches Earth's surface, most of it has either been scattered away by multiple reflections or been absorbed by the cloud, the dark base of the cloud being a lack of light entering the pupils of our eyes. We might then refer to the cloud as having a certain amount of transmittance or opacity. Physicists refer to this quantity at the surface of Earth as "the optical depth of the atmosphere," or a measure of the ratio of the incident light to transmitted light. The greater the optical depth, the smaller the amount of light that is transmitted to the surface (or the lower its intensity).

Why color and size matter

When we start considering very small scales in the atmosphere, light is also subject to further special, and quite magical, optical effects as it passes through the atmosphere, the end result being what gives our sky and its constituent clouds their color. The amount of scattering varies considerably with the wavelength of light and therefore its color, as well as the scattering particle, which may be a tiny molecule of air, a small aerosol, or a larger cloud droplet. Depending on the ratio of the size of the particle to the wavelength of the incoming light, all of the constituent colors of sunlight—all the colors of the rainbow—will be subject to different degrees of reflection, refraction, and diffraction when they interact with the constituents of the atmosphere (refraction being the bending of light and diffraction being the interference of light waves).

The net effect of all these optical effects is to bring an almost unique combination of color, light, and shade across the sky at any one time. For example, the sky appears blue because blue (and violet) light have short wavelengths, which are more easily scattered by molecules of air than longer wavelengths. Cloud droplets and ice crystals, however, which are larger than molecules of air and the wavelength of light, largely reflect white light. Alternatively, smoke or haze particles, which are even larger, may absorb blue light completely from the sky, leaving a brown or orange hue behind. In this way, familiar patterns of light and shade recur in the skyscapes of certain parts of the world at certain times of the day and year, often acting in concert with the evolving landscape beneath it.

A personal viewing experience

Finally, we must also factor in the influence of our own perception as a variable in our attempts to explain the light and color of the skies. These are largely subjective experiences, which vary from person to person, so can be difficult to fully interpret in a scientific sense. Added to this, many of us have differing qualities of vision—for example some are color-blind and some see colors differently from others—all of which adds to the diversity of our unique experience of the sky.

J.M.W. TURNER (1775–1851)

Turner is perhaps the quintessential Romantic painter, and was championed after his death by John Ruskin. Turner was famous for his turbulent landscapes and had a particular fascination with clouds, depicting every variety of cloud formation from sun-dappled gentle wisps to furious black storms. He was also impressively aware of the environmental degradation of his age: in *The Thames Above Waterloo Bridge*, he depicted the grimy exhaust fumes emitted by passing steamboats, although his equally famous *Rain, Steam, and Speed* presents a more positive vision of coming industrialization.

12. ***Inverary Pier, Loch Fyne: Morning* by J.M.W. Turner, ca. 1845**
Turner visited the Highlands of Scotland regularly. On this visit to Inbhir-Aoraidh (Inverary), located on the west coast, the weather appears to be reasonably calm, evidenced by the reflections from the waters of Loch Fìne (Fyne). Diffuse light dominates the scene, with patches of mist, fog, or low cloud (*Stratus fractus*) hanging close to the water; although a gap above reveals some blue sky and a hint of *Altocumulus*.

IRIDESCENCE

Iridescence is a colorful phenomenon that occasionally appears in short-lived mid- to high-level clouds. It arises from the diffraction of light—which, as we have learned already, is the interference of light waves—caused by tiny cloud droplets or ice crystals that are of a uniform size. From a physical viewpoint, diffraction (as opposed to refraction) becomes dominant when the size of the obstacle is of the same order of magnitude as the wavelength of the impacting wave. In this case, the cloud hydrometeors are so small that their diameters are directly comparable to those of the wavelength of light itself, so diffraction dominates.

The colors associated with the diffraction of sunlight—and occasionally moonlight too—tend to be more pastel or mother-of-pearl in hue than those of the rainbow: that is, they are perceived to have less saturation than bright or vivid colors. They are highly visually attractive, with soft hues and muted, delicate wavy repetitions. Adding to the spectacle, the colors frequently change position depending on the viewing or illumination angle, or due to movement of the cloud itself. We commonly see the same visual effect on the surface of soap bubbles, so next time you are doing the dishes, spare a thought for the beauty of both soap bubbles and clouds!

In the troposphere, *lenticularis* clouds (page 156) often show the best iridescence. This is because these mountain wave clouds develop and dissipate quickly as the air is constantly blowing through them, so there is not sufficient time for the embryonic cloud droplets to grow in size much beyond a micron or two. Iridescence is also more common in mid- and high-level clouds of the troposphere than in low-level clouds, as both cloud droplet and ice crystal sizes tend to be smaller the higher one goes in the atmosphere.

The best examples of iridescence, however, are reserved for the rare, stunning, and somewhat controversial polar stratospheric clouds, more commonly known as nacreous clouds. These form well above the troposphere, at altitudes of 12–25 miles (20–40 km) in the stratosphere. Their formation is linked to atmospheric pollution, climate change, and the hole in the ozone layer, so despite their stunning and ethereal beauty, they have something of a bad reputation (page 208).

Iridescent Clouds: Looking North from Cape Evans
by Edward Wilson, 1911

13. Iridescence is a beautiful mother-of-pearl phenomenon caused by the diffraction of light, which predominates when the size of cloud particles remains close to that of the wavelength of light. It is therefore most likely seen in cold, high clouds such as *Cirrocumulus lenticularis* (the cloud droplets or ice crystals not having much time to grow any larger). Iridescence is most spectacular in stratospheric nacreous clouds (page 208).

14. This is a spectacular capture by Wilson of nacreous clouds, forming over the Antarctic mountains just after the return of Sun in springtime (when the stratosphere is coldest). Nacreous clouds are usually lenticular in form, owing to the fact that the threshold temperature for their formation (-108°F/-78°C) is achieved more easily within the adiabatically cooled crests of each mountain wave.

13.

14.

SKY COLOR, SUNBEAMS, AND CREPUSCULAR RAYS

The principal optical effects of haze are to increase the scattering of light (or number of multiple reflections) along the line of sight, which changes its color. As we have learned already, when we see the sky as a shade of blue, it is because the shorter wavelengths of blue and violet light are scattered by tiny air molecules in preference to the longer wavelengths of orange and red. (This also explains why the sky appears jet black in outer space—there is nothing to scatter the sunlight). However, when it is hazy or dusty, with large aerosols abundant in the atmosphere that are much larger than individual molecules of air, the green, blue, and violet colors get scattered instead, or absorbed entirely, leaving behind the familiar yellows, oranges, browns, and reds that we associate with a hazy sky.

When the sky is both hazy and partly cloudy with light at a premium, sunlight shining through a gap in clouds may create the phenomenon commonly referred to as a "sunbeam." "Crepuscular rays," as they are known to meteorologists, appear to radiate outward from the Sun. This is simply an illusion caused by perspective; in reality the beams are always parallel with one another. More rarely spotted are anticrepuscular rays, which converge toward the antisolar point, directly opposite the Sun. Lunar "moonbeams" may also be spotted on occasion. Crepuscular rays are best observed early in the morning or late in the day when the Sun is low and shining through an extended path length of the atmosphere.

As well as being scattered, light waves may also become polarized; they begin to oscillate in a single direction, in contrast to unpolarized light, which vibrates in many or all directions (here we must consider light as a wave). We might expect light that has been heavily scattered and has undergone multiple reflections, such as light emanating from a diffuse cloudy or hazy sky, to be largely unpolarized. In contrast, smooth surfaces that sharply reflect light may produce concentrated beams of polarized light. This is why it is easier to view the brightly reflecting upper surfaces of *Cumulus* when wearing polarizing sunglasses, as they reduce the incoming polarized light, and consequent glare, considerably.

15.

15. ***Branch Hill Pond, Hampstead*** **by John Constable, ca. 1821–22** Constable captures a hazy sky to the west/southwest of London, with a characteristic emphasis on diffuse light and a weakened solar beam scattered by both clouds and aerosols. Mid-level clouds dominate the upper sky scene, with gray *Altostratus*, or possibly even orographically forced *Altocumulus lenticularis duplicatus* apparent (center to left). The wind at these levels is flowing from lower right to upper left (i.e., a northwesterly airflow). Surface conditions remain stable and anticyclonic, however.

In addition, it turns out that blue sky is most heavily polarized at ninety degrees to the Sun, with light being most heavily scattered in directions directly toward and away from it. This means that when the Sun lies in the south of sky, the sky will look bluest if you look in an easterly or westerly direction; conversely, it will look paler when looking directly toward the Sun or 180 degrees away from it. Furthermore, if you extend your arm and place your thumb over the Sun's disk (and can continue to look comfortably at your thumb while doing so), the air is likely to be very clean and contain very few aerosols, as there is little scatter other than blue light. Deep blue skies are also more commonly observed during periods of low relative humidity and reduced amounts of atmospheric water vapor; this is because any aerosols present are unlikely to become activated, thus reducing the chance of scattering.

In terms of skyscapes over the centuries, the occurrence of smoke, dust, haze, or activated aerosols have often provided the challenge, indeed the opportunity, for various art masters to attempt to capture not just meteorological realism, but also the subjective and expressionist emotions that these scenes and panoramas convey to us.

16. ***Calais Sands at Low Water, Poissards Gathering Bait***
by J.M.W. Turner, 1830
Here, we are looking west from Calais toward the setting Sun. Diffuse pale light and hazy skies dominate the scene, suggestive of sultry weather. It is likely warm, with thunder threatening, evidenced by a single *Cumulonimbus calvus* tower located in the distance. Prominent crepuscular rays (left of center, see page 80) emanate from other possible convection towers lying beyond the visible horizon.

6

5

4

3

“Turns them to shapes
and gives to airy nothing
A local habitation
and a name.”

William Shakespeare, *A Midsummer Night’s Dream*, Act 5, Scene 1

2

1

LOW CLOUD SPECIES

LOW CLOUD FAMILY TREE

Clouds are categorized according to the level at which their base forms, and the Low Cloud group includes all clouds whose bases form somewhere between Earth's surface and a height of 6,500 feet (2,000 m). There are five Low Cloud group genera: *Stratus*, *Nimbostratus*, *Stratocumulus*, *Cumulus*, and *Cumulonimbus*. Within each genus are species, characterized by specific physical characteristics or distinguishing behaviors; altogether there are thirteen exclusive species within the Low Cloud group and a further twenty-two shared and non-exclusive varieties, supplementary, and accessory clouds.

We shall work from the ground up, as befits clouds. Both *Stratus* and *Nimbostratus* are generally the lowest. They are the gray clouds that shroud the hilltops and bring fog to the coast; in cities they obscure the tops of the highest buildings. *Stratus* has two species: *nebulosus*, a uniform sheet; and *fractus*, a broken layer. *Stratus* is differentiated from *Nimbostratus* usually only by a steady precipitation that accompanies the latter. Somewhat confusingly, *Nimbostratus* is coded by WMO as a mid-level cloud with *Altostratus* (see page 140), although its base is always at a low level.

The other three Low Cloud genera are *Stratocumulus*, *Cumulus*, and *Cumulonimbus*. Being usually the direct products or byproducts of convection (thermals of rising air), a prerequisite of which is a well-mixed surface layer of air (or "boundary layer"), they tend to have somewhat higher bases than *Stratus* or *Nimbostratus*, usually starting somewhere between 1,000 and 5,000 feet (300 and 1,500 m) above ground level.

Stratocumulus is a rather dull, lumpy gray layer, and has five species: *stratiformis*, *lenticularis*, *castellanus*, *floccus*, and *volutus*.

Meanwhile, *Cumulus*, the cauliflower-esque heap, consists of four species: *congestus*, *mediocris*, *humilis*, and *fractus*. One key characteristic of well-developed *Cumulus* is the property of flat, level bases, which is a direct consequence of a well-mixed surface boundary layer.

The deepest and tallest of all clouds is *Cumulonimbus*, the King of Clouds; they are only categorized as "Low" because of where their base originates. In truth, they are the genuine masters of all three cloud levels, often soaring to the edge of the tropopause.

CLASSIFICATION OF LOW CLOUDS

GENERA	SPECIES, VARIETY, MOTHER CLOUD, OR GENERAL OBSERVATION	*	o
***Cumulonimbus* (Cb)**	*Cumulonimbus capillatus* (Cb cap)	C_L=9	
	Cumulonimbus calvus (Cb cal)	C_L=3	
***Cumulus* (Cu)**	*Cumulus congestus* (Cu con)	C_L=2	
	Cumulus mediocris (Cu med)	C_L=2	
	Cumulus humilis (Cu hum)	C_L=1	
	Cumulus fractus (Cu fra)	C_L=1	
	Cumulus and *Stratocumulus* with bases at different levels	C_L=8	
***Stratocumulus* (Sc)**	*Stratocumulus cumulogenitus* (Sc cugen)	C_L=4	
	Stratocumulus castellanus (Sc cas)	C_L=5	
	Stratocumulus lenticularis (Sc len)	C_L=5	
	Stratocumulus stratiformis (Sc str)	C_L=5	
***Stratus* (St)**	*Stratus nebulosus* (St neb)	C_L=6	
	Stratus fractus (St fra)	C_L=7	

* WMO code o International cloud symbol

WMO codes, abbreviations, and respective symbols for selected Low Clouds. For example, if a glaciated *Cumulonimbus* is observed, the code C_L=9 is recorded. However, not all clouds are represented—*lenticularis* being the most obvious species missing—as well as many other varieties, supplementary, and accessory clouds. Additionally, *Nimbostratus*, although almost always a low cloud, is coded at midlevels (for the benefit of weather forecasting and aviation purposes).

CUMULIFORM CLOUDS

When air is heated from below, it will rise of its own volition until it reaches a level of neutral buoyancy, in the same way that a beachball shoots upward after being held underwater (page 44), or a Montgolfier hot-air balloon rises when it is filled with heated air (page 45). Meteorologists designate this process of freely rising air "convection."

However, when Earth's surface is warmed enough to render the layer of air closest to the surface unstable, the air does not rise as a single slab or layer: instead, its ascent consists of many individual thermals or "updrafts," with typical dimensions of only tens to hundreds of feet in size, at least in regular and moderate convection. To compensate for the loss of air upward in the rising thermals, the hydrostatic balance of the atmosphere ensures that regions outside of the rising updrafts and their associated clouds will tend to descend slightly (but usually more slowly and across a wider area). A classic example of this is *Cumulus radiatus*, more popularly known as "cloud streets" (page 96), which consists of alternate parallel convective rolls of both rising and descending air.

Almost all low-level cumuliform clouds are formed by heating from below. Over land, this is usually by the Sun (or rarely by a volcano). Over oceans, it is the sea surface temperature that largely controls whether convection breaks out, which also depends on the temperature and moisture profile of the overlying airmass. Relatively warm inland water bodies, such as large lakes, can also produce strong convection when a deep mass of cold air moves over their surfaces, mainly in fall and early winter.

Mid- and high-level cumuliform clouds, however, usually do not owe their origin to heating from the surface of Earth. Instead, they owe their provenance to the movement of air currents associated with large-scale weather systems, for example when a cold layer of air is forced over a warmer one, leading to the outbreak of convection that is isolated from the surface, such as seen in *Altocumulus* or *Cirrocumulus castellanus* (pages 152–3).

The mighty *Cumulonimbus*, however, breaks all the rules: in its severest incarnation, individual updrafts may scale up to a mile (1.6 km) across, rising at speeds of 100 mph (160 km/h), exploding upward from low to mid- and high levels of the troposphere in just a few minutes.

1.

1. ***Study of Cumulus Clouds* by John Constable, 1822**
Today Constable's study would be more accurately titled *Study of Cumuliform Clouds* as the depicted cumuliform puffs and tufts ("ice cream castles in the air") more closely resemble those of the mid-level species *Altocumulus castellanus* (page 152) than its low-level cousin, *Cumulus*. This is no fault of Constable, it being less than 20 years since Howard first proposed *Cumulus*; it would take another half-century before mid-level "Alto" genera were accepted. Here, we are told that it is a warm September afternoon with a fresh east wind blowing. It is, therefore, not unreasonable to infer that air pressure is decreasing, with perhaps a chance of thunder ahead—i.e., the atmosphere is destabilizing, of which *Altocumulus castellanus* is a well-known precursor. Furthermore, streaks of high-level cirriform cloud (page 166) in the background indicate weather systems are possibly approaching. Meanwhile, abundant haze (lower right) from local, and also possibly foreign, sources indicate considerable scattering of light closest to the Sun—conditions often favored by Constable.

2.

CUMULUS (Cu)

2. ***Shinnecock Hills by*** **William Merritt Chase, ca. 1895**
Patches of "fair-weather cumulus" (*Cumulus humilis*, together with a few scraps of *fractus*) are apparent on this fine day, over Long Island, New York. The sunshine is strong, the air fresh and clean, both of which will have aided Chase to capture the vibrant, impressive colors of land and sky. Visibility is also excellent over the deep blue sea (left, center).

Cumulus
Cu

INDEX	
Genus	*Cumulus*
WMO codes	C_L=1, 2, 7, 8
Latin	"heap"
Species	*fractus* (fra) *humilis* (hum) *mediocris* (med) *congestus* (con)
Varieties	*radiatus* (ra)
Supplementary features	*virga* (vir) *praecipitatio* (pra) *arcus* (arc) *fluctus* (flu) *tuba* (tub)
Accessory clouds	*pileus* (pil) *velum* (vel) *pannus* (pan)
Appearance	Bright and fluffy
Frequency	Common

THE NATURE OF *CUMULUS*

The archetypal cloud that everyone thinks of first are *Cumulus*—from the early scribbles of a small child to the conceptual graphics of an executive's PowerPoint presentation as they explain the location of their servers and databases residing "in the cloud." The characteristic bubbly upper edge of a bright, newly formed *Cumulus*, protruding upward from a darker but mostly flat, homogenous base, is a ubiquitous symbol online, and can be found as an icon on most weather apps, regardless of what type of clouds are forecast to arrive.

Why are we biased toward *Cumulus* being the typical cloud? Perhaps it is because of its fractal nature, a common attribute throughout the natural world, whereby its shape and geometry is retained across scales, large and small, in a principle known as "self-similarity." One consequence of this is that, when viewing a *Cumulus* cloud from the window seat of an airplane, it can be disorientating, as it is difficult to judge its height, length, and scale without a good ground reference point. Or maybe it is because *Cumulus* clouds, when forming over land, are largely diurnal like us, rather than nocturnal, owing their origin to the thermals created by the land surface after being heated by morning sunshine.

How *Cumulus* arises

Cumulus clouds form when there is instability in the lowest layers of the atmosphere. Instability means that air has a tendency to rise of its own volition, due to buoyancy (page 44)—just like Montgolfier's hot-air balloon, or the beachball that always shoots back up to the water's surface. In both cases, the air is simply attempting to return to a level of neutral buoyancy.

Instability can be created in many ways in the atmosphere, most commonly when the Sun heats Earth's surface in the morning, producing a thermal of air that is soon warmer than the air lying above it. It will then start to rise due to its own natural buoyancy.

On a broader scale, the "advection," or movement of air masses, can also cause instability to break out: for example when a cold, dry air mass (such as that following a cold front) moves over a warm surface (such as a warm ocean surface). This instability will continue, both day and night, as long as the supply of cold air is maintained and the ocean water temperature remains warm enough for evaporation to create moist rising thermals. Such weather situations are common over the Pacific and North Atlantic oceans when cold air drains off nearby continents (see illustration page 44).

3.

3. ***At the Seaside* by William Merrit Chase, 1892**
In a style more realistic than Constable's, Chase painted this scene on a fine day at Long Island, New York. Shallow fair-weather *Cumulus humilis* lie almost overhead but have spread out and flattened into patches of *Stratocumulus perlucidus* in the distance, indicating the likely presence of anticyclonic inversion. The yacht and choppy waves suggest that the umbrellas may be acting more as windbreaks than sunshades.

THE GEOMETRY OF *CUMULUS*

There are four species of *Cumulus*. Listed in increasing order of sheer volume or bulk, they are *fractus*, *humilis*, *mediocris*, and *congestus*.

Cumulus fractus

The *fractus* species of *Cumulus* consists of well-separated ragged or shredded wisps of cloud. *Cumulus fractus* is typically formed over land early in the morning by the first rising thermals of air, saturating only temporarily as they rise and mix with drier air. They quickly evaporate after a matter of seconds, only to be replaced by others, or to gradually give way to more powerful thermals as daytime convection becomes stronger.

Cumulus humilis

Representing the next stage in the development of the *Cumulus* family, the cloud cells of *Cumulus humilis* remain relatively small and well separated, and are horizontally wider than they are tall. On a pleasant summer's afternoon, they can often be observed to exist in a quasi-harmonious state of moderate ascent, gentle overturning, decay by evaporation, and rebirth nearby, all over the course of just 5 or 10 minutes—the whole process being best revealed using timelapse photography. Due to their limited vertical extent and short lifetime, they produce no precipitation, which is why *humilis* is often referred to as "fair-weather cumulus."

Cumulus mediocris

On the other hand, *Cumulus mediocris* is the somewhat livelier and more energetic bigger sibling of *humilis*. With more powerful thermals giving rise to moderate turbulence within the cloud, the *mediocris* species is about as wide as it is tall, leading to a longer lifetime (perhaps 20–40 minutes). Again, it tends to exist in a state of near equilibrium on a relatively fair day: ascending more quickly than *humilis*, overturning, evaporating, and eventually being reborn nearby. Occasionally, however, a few *mediocris* cells may display "teenage tendencies," and start to break out from the regular pattern, growing somewhat bigger than their neighbors and producing a little precipitation. Interestingly, due to the different cloud condensation nuclei contained in maritime *Cumulus* compared with continental *Cumulus* (page 59), the maritime version of *mediocris* is more likely to produce precipitation, but usually just a brief passing shower.

4.

5.

6.

4. ***Cloud Study* by Knud Baade, 1838**
A somewhat idealistic portrayal of *Cumulus fractus*: scraggy, lacking any structure, and illuminated by a very low Norwegian Sun near twilight. It is possible these clouds may be leftover remains (*pannus*) of earlier large *Cumulus* or *Cumulonimbus* showers.

5. ***Landscape with Cumulus Clouds* by Andreas Schelfhout, ca. 1839**
Cumulus fractus and *mediocris* turrets on the verge of building into larger *Cumulus congestus*, tilted by an increasing wind with height from left to right, and illuminated by a low Sun. At higher altitudes are some patches of white cirriform (ice) cloud, set against the backdrop of fine blue sky. There may be a weather front approaching within the next 12–24 hours.

6. ***The Calm Sea* by Gustave Courbet, 1869**
Although the title states "calm," neither the ocean nor especially the clouds are indicative of tranquility. Sailing boats and a ruffled sea surface with some white horses suggests a fresh breeze, while the unusually low clouds indicate a high degree of instability in the atmospheric layers closest to the sea surface. The juxtaposition of both white and gray rising wisps of cloud appear out of place meteorologically. Perhaps Courbet added the clouds later, for effect?

7.

8.

Cumulus congestus

Meanwhile, *Cumulus congestus* is the adolescent of the family, growing powerfully, rapidly, and valiantly upward, but not yet mature—that status is reserved solely for the separate (and mighty) genus of *Cumulonimbus* (page 126). *Congestus* is easily identified by soaring towers of bubbling *Cumulus*. The cloud is taller than it is wide, stunningly bright on its sunlit, cauliflower-like, fractal upper surface, but with a dark and threatening base bringing the risk of localized heavy showers (supplementary features *praecipitatio* or *virga*), hail, and gusty winds. Rarely, it might even produce an *arcus* cloud (page 202) or, more rarely, *tuba* (page 206).

Cloud streets

The *radiatus* variety of *Cumulus* refers to regular lines or "streets" of the cloud that align themselves parallel with the prevailing airflow (wind) direction. Looking directly upwind, the cloud streets therefore appear to radiate from a single point near to, or just below, the horizon. Each cloud street is typically separated by a few miles of clear air. The clouds themselves sit atop the upper part of a system of horizontal convective rolls, all of which advects continuously downwind (they are, essentially, long cylinders of counter-rotating air oriented parallel to one another as well as being aligned along with the wind). *Radiatus* is most commonly expressed in the species *Cumulus mediocris*, but may also occur with *humilis* and occasionally *congestus*. The *radiatus* variety is also common to *Stratocumulus*, as well as to *Altostratus*, *Altocumulus*, and *Cirrus*, although with different formation mechanisms.

Due to the large liquid water content of *Cumulus* clouds, and their bright and reflective upper surfaces, they create big variations of light and shade on Earth as they pass overhead. This is perhaps one of the reasons why they are so photogenic—and when the Sun is low, they can produce stunning sunbeams, or crepuscular rays (page 80), and sometimes even anticrepuscular rays.

PILEUS, *VELUM*, AND *PANNUS*

A deep and well-developed *Cumulus* or a regular *Cumulonimbus* may sport the accessory clouds *pileus*, *velum*, and *pannus*. *Pileus* and *velum* are similar to "mountain cap" clouds (page 210)—that is, they are laminar and lenticular-style clouds through which the convective cloud continues to grow stridently and independently, behaving as though it were as a mountain top itself, and pushing the ambient airflow around to its side (producing an extensive skirt of *velum*) or over its top (forming a hood of *pileus*). Meanwhile, *pannus* (scud) is common to all precipitating clouds, and consists of ragged wisps and shreds of low cloud or "scud" lying below the general cloud base, formed either by precipitation evaporating as it falls, or by local turbulence.

7. ***Cows Crossing a Ford* by Jules Dupré, 1836**
Dazzling white tops of rows of *Cumulus mediocris radiatus*, on the verge of building into *Cumulus congestus*. Some patches of *pannus* or *fractus* are also present. Heavy showers may break out.

8. ***Landscape with Cattle at Limousin* by Jules Dupré, 1837**
The skyscape forms the principal part of this depiction of a rural scene in France. There's likely a moderate or fresh westerly breeze blowing in from the Atlantic (recent rain has kept the skies clean and the landscape green). In the distance we can see at least six parallel rows of "cloud streets" (*Cumulus mediocris radiatus*), aligned parallel with the wind. Some of these have spread into patches of *Stratocumulus* closer to the artist's location.

FREDERIC EDWIN CHURCH
(1826–1900)

Along with his teacher, Thomas Cole, and the rest of the Hudson River School of New York, Church specialized in beautifully realized landscapes depicting the scale and majesty of the American natural world. Although influenced by the idealism of the Romantics, the accurate and exquisite brushwork of Church's paintings offers the viewer an invaluable insight into what the pre-industrial North American landscape might have looked like. It is a landscape that was already rapidly disappearing in Church's time.

FLAT BASES

The *Cumulus* family owes its origin to the turbulent, well-mixed, lowest layer of the troposphere; this means that the air mass characteristics of a fixed reduction of air temperature with height (known to meteorologists as the "adiabatic lapse rate") and fairly uniform amounts of moisture remain constant throughout the sub-cloud region. Barring any significant changes to the heat and moisture profile of the airstream during the daytime hours, this means that the air being lifted into the *Cumulus* clouds will tend to saturate at exactly the same altitude. This produces the characteristic and frequently observed feature of *Cumulus* clouds—their level bases. So, even though the environment within each *Cumulus* cloud is highly turbulent, a remarkably uniform horizontal cloud base, lying at constant elevation, can usually be seen as the clouds drift across the landscape, lending harmony to an otherwise chaotic process.

It was this "level base" property of *Cumulus* clouds that Ruskin used in *Modern Painters* (1857), along with those of layer clouds such as *Altostratus*, *Altocumulus*, and *Cirrocumulus*, in his attempts to apply the rules of perspective and geometry to the natural appearance of cloud formation (page 39).

Perturbed bases

In contrast, on the rarer occasions when *Cumulus* clouds do not have flat bases and are uneven, this tells us something valuable about the lower atmosphere—either that the air is not well mixed, or that there is an air mass boundary nearby (such as a weather front), or something local is perturbing the airflow (for example a mountain). Alternately, the cloud itself may be taking charge, effectively generating its own weather, as evidenced by supplementary or accessory cloud features such as *virga* or *pannus* created by local turbulence beneath the cloud base. Much more rarely, and usually associated with *Cumulonimbus*, a *tuba* (funnel cloud) may appear in the base, indicating cloud rotation. *Murus* and *cauda* (accessory clouds of *Cumulonimbus*, page 126) are direct evidence that the cloud itself, rather than the ambient airflow, is now fully controlling the environment. The cloud is now generating its own, often severe, weather.

9. ***Cloud Study* by Frederic Edwin Church, 1871**
Puffy turrets of *Cumulus congestus radiatus* align neatly in a cloud street (*radiatus*). The prominent level base indicates a well-mixed turbulent and adiabatic environment in the air below the cloud. The surface wind is likely blowing from right to left, as the clouds get deeper downstream to the left. Fresh mid-level breezes appear to be tilting the close turrets in alternate directions.

10. ***Nightfall near Olana* by Frederic Edwin Church, 1872**
We can't quite see the base of this towering *Cumulonimbus calvus*, as it is after sunset on the surface, with the last of the Sun's rays illuminating the cloud top only. Ragged low-level scud (*pannus*) ahead of the cloud are suggestive of a moist environment with rain nearby.

9.

10.

ANABATIC *CUMULUS*

11. ***Torrent in the Highlands*** **by Gustave Doré, 1881**
Scattered *Cumulus fractus* forms over sunlit high ground. Doré likely painted this scene early in the morning, as the mountain-valley air circulation system is not yet fully established; valley mist (or more likely smoke from peat fires, as indicated by its distinctive blue color) remains lying in the shadow of the largest mountain peak, as yet largely undisturbed.

On a warm spring or summer's day, you will have likely noticed that *Cumulus* clouds prefer to form initially over hill and mountain tops. The same thing happens over oceanic islands—by day *Cumulus* forms quickly and remains anchored quite persistently over isolated islands and near coastal zones. Over the centuries, mariners and explorers have used this knowledge as a vital clue when searching for land that is hidden below the horizon.

The reason why this happens is straightforward. At a specific altitude, rising thermals lower the air pressure above a mountain's surface more than they do over any adjacent valleys. The air pressure is lowered by the increase in temperature, because warm air is less

dense than cold air. The same process operates over islands, the air becoming warmer directly over the island than at the same height over the surrounding sea. This leads to a horizontal pressure gradient between the two zones, and since air mostly flows horizontally, a temporary air circulation system is established.

"Anabatic" breezes (from the Greek "to ascend") blow upslope toward the mountain summit, producing *Cumulus* clouds. A corresponding descending current of air returns over the valley, leading to clear skies. The complete mechanism is known as a "mountain-valley air circulation system." Around an oceanic island, the same process operates; the afternoon sea breeze blows gently onshore in a convergent fashion around the island, with a corresponding airflow returning back to sea at an altitude of a few thousand feet.

12.

STRATIFORM CLOUDS

12. ***Greifswald in Moonlight*** **by Caspar David Friedrich, 1817**
One gets the feeling of a cold, gray winter's evening in Friedrich's home town, Greifswald. As the Moon's disk is discernible behind the spire of St. Nicholas' Cathedral, the cloud's variety is *translucidus*. Owing to its gray appearance, the principal cloud layer is probably *Altostratus*, with a hint of *undulatus* (upper center). In the distance, there appear to be some gaps (variety *perlucidus*). A thin surface mist or haze also seems to permeate most of the town's surroundings.

HOW STRATIFORM CLOUDS FORM

Cloud layers are often the result of ascending currents of air that have had to spread sideways, with any strong vertical movements being checked by a "capping inversion." The inversion is usually an invisible layer of warmer and drier air with enhanced stability lying directly above rising thermals. The result is that the buoyant energy of rising thermals is quickly suppressed, with the momentum instead being transferred sideways.

STRATIFORM CLOUDS

On social media, many of us tend to bias our profiles by highlighting the most important and colorful moments in our lives and excluding our more mundane activities. And so it is with clouds. The big, the beautiful, and the vertically dramatic get media attention and tend to be overexposed online (think *Cumulonimbus*, tornadoes, and storms), while the far more common and seemingly banal layer or "stratiform" clouds, such as *Nimbostratus*, *Stratus*, or *Stratocumulus*, generally do not provoke such enthusiasm.

Layers, both of cloud and of air, are ubiquitous in the atmosphere and are not as boring as they may appear, despite the fact that layer clouds are by far the most common type of cloud in the atmosphere. To start with, layer clouds such as *Stratus* can tell us something about the state of the atmosphere. We can usually take a guess at their provenance—where and how they have formed, which is often as a younger and more buoyant cloud—and they may also tell us what the prognosis may be.

Sometimes, flat layers of clouds are not even quite that—they are sloped, but with gentle and largely imperceptible gradients, many times less steep than mainline train gradients. Within them, the air may be rising only a few millimeters per second. From the ground, such inclines are largely indistinguishable from the horizon, and so we perceive them as flat layers. At other times, layers of cloud may be stacked (variety *duplicatus*) upon one another, with each layer progressively lowering as a weather front approaches. More often, however, they are formed by the horizontal spreading out of previous vertically ascending air currents (see box).

In these ways, *Stratus* or *Stratocumulus* clouds are formed at low levels. At midlevels, *Altostratus* and *Altocumulus* may be arranged in the same way; the same is also true for cirriform clouds at high levels.

13. ***Purple Clouds*** **by John Ruskin, 1868**
Stratiform or layered clouds are most commonly observed near dawn, for the simple reason that the Sun has yet to warm the surface to produce thermals of rising air. Even so, a little convection can still occur on the upper surface of cloud layers overnight, due to cloud-top cooling by infrared radiation, sometimes forming *Altocumulus* or *Cirrocumulus*. Here we see scattered tufts of what appear to be *Altocumulus* lit up brightly by the first rays of the rising Sun (located just off bottom left). The other cloud layers are probably *Altostratus* or thick cirriform clouds; some streaks of *virga* are also present (top center and right).

13.

Stratus
St

INDEX	
Genus	*Stratus*
WMO codes	C_L=6, 7
Latin	"layer"
Species	*nebulosus* (neb) *fractus* (fra)
Varieties	*opacus* (op) *translucidus* (tr) *undulatus* (un)
Supplementary features	*praecipitatio* (pra) *fluctus* (flu)
Accessory clouds	None
Appearance	Dull, low, gray
Frequency	Common

THE NATURE OF *STRATUS*

Stratus ("layer" in Latin) is the lowest of the layer clouds, being a rather monotone, featureless blanket of gray cloud whose base usually lies less than 1,000 feet (300 m) above our heads. It is a common visitor on chilly winter days in the damp climates of the middle latitudes. *Stratus* is strictly a low-level cloud; if the cloud extends into midlevels or higher, and if rain or snow is falling from it, it is then described by the separate genus *Nimbostratus* (page 122).

Long-lived *Stratus*

Lying close to the surface, *Stratus* is one of the longest-living clouds. This is because winds in the lower troposphere are generally much lighter than at higher altitudes, and consequently the mixing of air, especially during slack high pressure (anticyclonic) situations in winter, is limited or even non-existent. Widespread layers of near-surface *Stratus* can therefore persist over continental regions for days on end in winter, or even for weeks, especially if the low cloud is enclosed by mountains and well protected from the encroachment of other air masses. In these situations, which are often associated with poor air quality in industrial areas, cloud-top radiational cooling during long winter nights maintains a condition of near equilibrium, with weak winter sunshine failing to disperse it during the short winter days.

Stratus is also frequently found over the ocean, often in association with its close cousin *Stratocumulus* (page 118). These extensive layers of maritime stratiform cloud, as revealed by analyses of satellite images over the past half-century, are ubiquitous. They are a consequence of persistent cloud-top cooling (by infrared radiation to space), combined with a gentle but steady evaporation, all day, every day, from the surface of the ocean. This, in turn, leads to a persistently humid near-surface environment. In the presence of a low-level tropospheric temperature inversion—as is often found under the subtropical "Hadley Cells"—any rising cloudy convection soon spreads out into layers of *Stratus* or *Stratocumulus*. Hydrostatic balance (page 52) in the atmosphere also helps to continuously maintain a steady state of equilibrium, preventing any sudden vertical movements of air, which might disturb the cloud.

Stratus can also be formed by the gentle descent of cold, saturated air, when it pools in sheltered valleys overnight, or on plains, after draining from surrounding cold mountains and hillsides: the cold pool of air eventually reaching a depth of up to 1,000 feet (300 m) or so. Such *Stratus* is common across large parts of central Europe during spells of anticyclonic weather in the winter, and the mountain tops (for example the Massif Central or the Alps) can usually be seen rising dramatically above a sea of cloud, looking like Greenland nunataks (glacial islands).

14.

15.

14. ***Studies of Alpine Peaks* by John Ruskin, 1846**
In this apparent montage of Alpine summits, the mountain peaks rise majestically above a sea of *Stratus* below, as if they were Greenland nunataks protruding above the ice cap. The wispy streaks at mountain peak level may represent patches of *Altostratus* (page 140) or even *Cirrus radiatus* (page 171) for effect. Alternatively, they may be parts of banner clouds (page 210) or just wind-blown snow.

15. ***Champagnole* by John Ruskin, 1846**
Inky gray (with a hint of indigo), *Stratus* or *Stratocumulus* spreads from left to right over the French Jura, its color echoed in the hues of nearby hills. A deterioration in the weather may be expected, with cloud bases lowering as the cloud advances. Traces of mist or haze are discernible in the valley below, with a patch of *silvagenitus* (page 194) forming over the conifer trees, indicating recent rainfall.

In this meteorological situation, a strong temperature inversion, coupled with and enhanced by radiational cooling (at infrared wavelengths) at the top of the cloud, maintains a persistent and stable equilibrium. With reverse convection operating, the cold air at the top of the cloud overturns and sinks down into the cloud due to its increased density caused by cooling.

THE GEOMETRY OF *STRATUS*

Stratus has two species: *nebulosus* and *fractus*. *Nebulosus* is used to describe a uniform, featureless, blurry, monotone gray layer, and is the most common species of *Stratus*: one which we might encounter on a dull, cold, dry winter's day. It is best observed from above—from an airplane or a mountain summit—from where it occasionally yields a spectacular glory, or "Brocken specter."

In contrast, and as with *Cumulus* (page 90), the *fractus* species of *Stratus* consists of wisps of low, ragged cloud, initiated by turbulence or evaporation close to Earth's surface, forming under the more general cloud base at an altitude of only a few hundred feet (100–200 meters). *Stratus fractus* can commonly be observed over forest canopies shortly after the cessation of rain, when it is given the special "mother cloud" name *silvagenitus* (page 194).

Stratus also has three varieties (*opacus*, *undulatus*, and *translucidus*) and two supplementary features (*praecipitatio* and *fluctus*). These names are fairly self-explanatory; *opacus* is opaque to the direct rays of the Sun (its disk remains hidden from view), whereas *translucidus* lets some light through (the outline of the Sun's disk is visible). The *undulatus* variety refers to visible wave-like patterns in the cloud; transient waves or billows can be set up when the wind starts to increase just above the top surface of the cloud, or large waves may be set up when the air flows over or around nearby hills or mountains. If the wind speed increases significantly within the cloud layer, the whole deck is likely to break up and dissipate before too long.

Alternately, if small breaking waves are seen on the upper surface of the *Stratus* deck, these may develop into transient, beautiful breaking waves known as *fluctus* (page 198). Finally, *praecipitatio* is appended when light drizzle or snow flurries fall from *Stratus*—this occasionally happens in moderately turbulent *Stratus*, mostly in maritime or coastal areas.

Visibility and light levels

Stratus is a low-level cloud with its base lying close to the ground in conditions of high relative humidity, so when it is present, visibility is often heavily restricted at Earth's surface: a characteristic it shares with *Nimbostratus* and occasionally *Stratocumulus*. This is not necessarily the case for other stratiform clouds such as *Altostratus* or *Cirrostratus*, as they may lie above drier low-level layers of air.

Stratus is rich in water content, so it attenuates the Sun's rays considerably as they pass through the cloud, lending a dull and dreary character to conditions on Earth. However, the average thickness of a deck of *Stratus nebulosus* is usually only 600–1,200 feet (200–400 m), so it lets through more light than does, for example, *Nimbostratus* because the latter sometimes spans mid- and high cloud levels as well.

16.

16. ***Fishing Boats Becalmed off le Havre* by J.M.W. Turner, undated**
What looks like a featureless gray layer of *Stratus nebulosus* lies monotonously over the sea, with a small break of brighter skies apparent in the distance. An undisturbed sea surface and a becalmed boat in the foreground indicate light winds and anticyclonic conditions.

THE OSCILLATION OF *STRATUS*

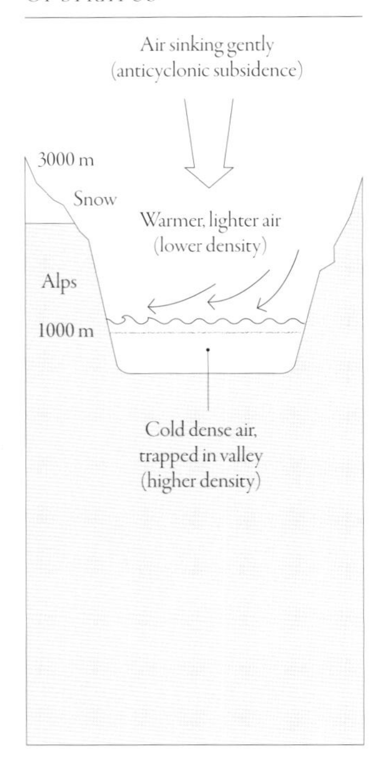

Schematic depiction of *Stratus* trapped in a Caucuses valley
If stimulated by a little wind on its upper surface, a regular, harmonic oscillation will be set up. The period of oscillation—the ti me between successive high or low "tides"—depends only on the depth of the trapped cold air, the contrast in density between it and the warmer air lying above the trapping inversion, and the width of the valley.

SEICHE SLOSHING

At first glance, a gray, featureless, overcast sky of *Stratus* does not generally evoke feelings of excitement or elation—more commonly we might describe it using adjectives such as depressing, boring, or monotonous. Equally, thick fog can provoke feelings of enclosure and disorientation; it may also be a nuisance and is a danger when traveling. However, if you are perceptive and have plenty of free time on your hands, you might just be able to perceive some of the hidden but fascinating behavior of *Stratus*.

Like the tea in your teacup, water in the bath, or tides in the ocean, fluids, even when disturbed only lightly, tend to slosh about and oscillate back and forth in a periodic and harmonic manner, hardly losing any energy in the process. The factor disturbing the tea in your teacup is just your shaky hand. In the bath it is quite easy to find the resonance and splash the waves over the side: any child knows that. In the ocean, the gravity of the Sun and Moon, the wind, and the local coastline all contribute to how steeply the tide flows.

How oscillation arises

It is no different for any other fluid when influenced by external forces, including surface fog or low *Stratus* cloud trapped in a valley or lying across a plain between hill ranges during an anticyclonic spell of winter weather. With time, and if stimulated by a little wind on its upper surface, it too will begin to oscillate in a regular, harmonic manner. The period of oscillation—the time period between successive high or low "tides"—depends only on the depth of the trapped colder, denser air, the contrast in density between it and the warmer, less dense air lying above it, and the width of the valley or plain within which it is contained.

Timelapse animations of fog and *Stratus* taken by webcam images on Alpine summits in Switzerland confirm these regular oscillations. Known as "seiches," a word borrowed from oscillations of the water surface of nearby Lake Geneva, *Stratus* in the Swiss valleys has wave periods that generally range from 7 to 20 minutes.

17. ***From Mleta to Gudauri* by Ivan Konstantinovich Aivazovsky, 1868**
Here the valley fog lies in a narrow, deep Caucasus valley. Like the landscape, the meteorology is exaggerated—valley *Stratus* usually has a smoother upper surface. The patches of cloud in the distance surrounding the snow-capped mountain peak may be wisps of transient *Cumulus fractus*.

18. ***Fluelen Morning (Looking Towards the Lake)* by J.M.W. Turner, 1845**
Seiches were first observed as periodic oscillations in the water surface of Lake Geneva, Switzerland.

17.

18.

19.

STRATOCUMULUS (Sc)

19. ***Landscape with Clouds*** **by John Constable, 1822**
This is an excellent portrayal of *Stratocumulus perlucidus cumulogenitus*, or *Stratocumulus* which has recently formed from the spreading out of the moist updrafts of *Cumulus*. The key element that distinguishes it as *Stratocumulus* is that the cloud cells are joined together and are not separate, with just occasional breaks between them (variety *perlucidus*, the brightest *Stratocumulus* formation).

Stratocumulus
Sc

INDEX	
Genus	*Stratocumulus*
WMO codes	C_L=4 C_L=5 C_L=8
Latin	"heaped layer"
Species	*stratiformis* *lenticularis* *castellanus* *floccus* *volutus*
Varieties	*translucidus* *perlucidus* *opacus* *duplicatus* *undulatus* *radiatus* *lacunosus*
Supplementary clouds	*virga* *mamma* *praecipitatio* *fluctus* *asperitas* *cavum*
Accessory clouds	None
Appearance	Dull, gray, lumpy; overcast or with occasional breaks
Frequency	Very common

THE NATURE OF *STRATOCUMULUS*

Stratocumulus (from *stratus* meaning "layer" and *cumulus* meaning "heap") is the most common low-level cloud. It is ubiquitous in middle latitude coastal climates, as well as over some continental interiors during winter. It is also observed frequently over oceans.

As its name suggests, *Stratocumulus* is a mixture between *Stratus* and *Cumulus*. In reality, it is neither, yet possesses the less appealing qualities of both these genera, making it a good candidate for the cloud that always disappoints. Lacking both the bright, buoyant youthfulness of *Cumulus*, and the quiet, reserved nonchalance of *Stratus*, it is the cloud that has spoiled many a hopeful summer's day. It may block out the Sun and make it feel chilly, through the sudden building up and spreading out of *Stratocumulus cumulogenitus* (pages 112–13), or seemingly refuse to dissolve and break up on what might otherwise have been a fine day.

As with *Stratus*, *Stratocumulus* usually lies as a single deck of low-level cloud, but it is lumpier and more heaped, and lacks the more relaxed, laminar character of *Stratus*. *Stratocumulus* also lacks the strong vertical motions of a fresh, bubbling *Cumulus* cloud—instead, there are more gentle vertical motions that are, in opposition to *Cumulus*, largely generated from the cloud top downward. Also, *Stratocumulus* prefers a more sluggish approach, often lasting for hours: although, like *Cumulus*, it does continuously regenerate itself through overturning, but at a gentle rate. Despite these significant differences, sooner or later both *Cumulus* and *Stratus* may morph into a more depressing modification, such as *Stratocumulus cumulogenitus* (pages 112–13) or *Stratocumulus stratomutatus*.

Five different species of *Stratocumulus* are recognized—*stratiformis*, *lenticularis*, *castellanus*, *floccus*, and *volutus*—the joint-highest tally of any cloud, sharing the statistic with its mid-level cousin *Altocumulus*. The species *stratiformis* is the most common, with the clue to its nature being in the name; *Stratocumulus stratiformis* describes an extensive, dappled, slightly heaped, uniform gray layer of cloud, without any clear breaks.

In common with *stratiformis*, the species *lenticularis*, *castellanus*, and *floccus* are also shared with the mid- and high-level cloud genera *Altocumulus* and *Cirrocumulus* (see pages 148 and 180, respectively). The new cloud species *volutus* is also common to both *Stratocumulus* and *Altocumulus* (page 198).

Unlike *Cumulus*, which is sustained by thermals rising from Earth's surface, *Stratocumulus* is powered instead by a much gentler circulation that arises from cooling by infrared radiation of the top surface of the cloud.

20.

21.

22.

23.

***Evening Landscape After Rain* by John Constable, 1821**

20. From Constable's painting we can conclude that the lower troposphere is conditionally unstable to slight convection, but the prognosis appears good. Perhaps a cold front with rain passed through the region earlier in the day?

21. *Detail:* In the clear skies beckoning, there are abundant instances of the wave cloud *Stratocumulus lenticularis* (page 158), an indicator of atmospheric stability, and which are usually initiated by mountains or hills upwind.

22. *Detail:* Above these clouds, the sky appears largely clear at mid- and upper levels; the air here is therefore dry and possibly descending due to high pressure.

23. *Detail:* What looks like a shallow *Cumulus congestus* tower rises and punches through this layer of stability, before tumbling over and modifying into *Stratocumulus* downstream. Another of these towers lies directly overhead (in main painting).

THE GEOMETRY OF *STRATOCUMULUS*

There are seven varieties of *Stratocumulus*, the greatest number of any cloud other than *Altocumulus*, with which it shares this metric. For *Stratocumulus*, these varieties are *translucidus*, *perlucidus*, *opacus*, *duplicatus*, *undulatus*, *radiatus*, and *lacunosus*.

The first three refer explicitly to the transmission (or suppression) of light through or around the cloud, and their names are largely self-explanatory. *Translucidus* describes a specific translucent variety of the cloud, whereby the disk of the Sun or Moon can still be determined through the individual cloud elements, and which remain fairly bright to the naked eye. *Perlucidus* is used when the cloud deck has edges and gaps, between which the sky above can be clearly seen. Conversely, *opacus* describes a completely opaque layer of the cloud through which neither the Sun nor Moon can be determined, appearing rather dull and gray to an observer on the ground.

The remaining four varieties—*duplicatus*, *undulatus*, *radiatus*, and *lacunosus*—all refer to different structural formations of *Stratocumulus*. Again, the names used are mostly self-explanatory.

Duplicatus describes *Stratocumulus* when it occurs in vertically repeated layers, as if it were duplicated over two or more levels, quite close to one another. *Undulatus* is used when oscillatory or undulating patterns are apparent in distinct wave-like fashion, such as stripes or billows (page 154) aligned *perpendicular* to the airflow, and which migrate with the cloud. As with *Cumulus*, and in direct contrast with the *undulatus* variety, the *radiatus* variety of *Stratocumulus* refers to lines or "streets" in the cloud that are aligned *parallel* to the prevailing airflow.

Finally, *lacunosus* refers to a special arrangement of small clear spaces resembling a honeycomb, around which the individual cloud elements are arranged—it consists mostly of clear sky.

Stratocumulus also has six supplementary clouds associated with it: *virga*, *praecipitatio*, *mamma*, *fluctus*, *asperitas*, and *cavum*. As was the case with *Cumulus* and *Nimbostratus* (page 124), *virga* refers to trails of precipitation falling from the base of the cloud that do not reach the ground; when they do, *praecipitatio* is used instead. *Mamma* are best displayed on the underside of a *Cumulonimbus* anvil (page 130), but can also occur on the underside of *Stratocumulus cumulogenitus* (pages 112–13). *Fluctus*, *asperitas*, and *cavum* are all new supplementary clouds; these are discussed in detail in Chapter 6.

24.

25.

24. ***Stone Age Mound*** **by Carl Gustav Carus, ca. 1820**
Despite the artist trying to turn our gaze toward the emotive effect of the full moon lying directly over the remains of the Neolithic circle, from a meteorological viewpoint, the cloud species and its varieties are easily defined as *Stratocumulus translucidus perlucidus*.

25. ***Sunset*** **by Frederic Edwin Church, 1850–80**
The long street of *Stratocumulus radiatus* owes its origin to the crest of the hills, and the paler high-level deck of *Cirrocumulus* might also be orographically forced. The vivid sunset hues and crystal-clear skies above resonate with the frontier landscape.

THE GEOMETRY OF MARINE *STRATOCUMULUS*

When viewed from a global perspective, *Stratocumulus* is by far the most common and most widespread low-level cloud. This is because vast sheets of it are to be found lying over the oceans, drifting slowly around the huge high-pressure subtropical zones, seemingly for days on end.

It was not until after the advent of the satellite era in the late twentieth century that scientists were able to confirm that *Stratocumulus* covers as much as one-fifth of Earth's oceans at any given time. Outside the tropics and subtropics, *Stratocumulus* is also frequently encountered in the climates of the middle latitudes and polar regions of both hemispheres, often hidden under other higher cloud decks in cyclonic systems or lying persistently over the polar oceans during summertime.

If we adopt the satellite's perspective for a moment, looking down on Earth from a considerable distance above, we would be able to recognize at least three more types of *Stratocumulus* over the oceans, which appear on a much larger scale, and are termed "open-cell," "closed-cell," and "actinoform" types.

Shape and form

Open-cell *Stratocumulus* consists of regularly patterned rings of broken cloud surrounding areas of clear sky. They are roughly polygonal—almost hexagonal—in arrangement. Somewhat resembling the honeycomb structure of *lacunosus* (page 116), but occurring on a much larger scale, the clear areas in between the cloud usually range from 3–30 miles (5–50 km) across. Although the surface weather is usually fine or fair beneath the open cells of *Stratocumulus*, the cloudy zones sometimes produce light drizzle or a brief shower. Identical open-celled cloud structures also appear in mid-latitude and polar *Cumulonimbus*, which contain heavier showers (page 132). The patterned structure of such formations is usually only visible from space.

Conversely, closed-cell *Stratocumulus*, as its name implies, completely covers the sky, blocking out direct sunshine and reflecting most of it back to space—as a result, the cloud appears bright from above but dull from below. As the sky is uniformly covered by gray cloud, it is easy to recognize this type from the ground—it is usually classified as *Stratocumulus stratiformis opacus*. The individual cloudlet cells are visible from the ground, appearing as regular dapples on the surface of an otherwise gray, uneventful layer of mostly uniform cloud. Unlike the open-celled version, there is usually little or no precipitation associated with the subtropical maritime version of closed-cell *Stratocumulus*.

Actinoform cloud structures (from the Greek for "ray") have only recently been discovered through the use of timelapse

26.

animation of high-resolution NASA satellite images. The adjective "actinoform" does not describe any particular cloud species, feature, or variety—rather it describes the attractive but unusual, large-scale "rose" or "star-shaped" structure of *Stratocumulus*, which may span 60–180 miles (100–300 km) across, and occurs only over the open ocean. Actinoform cloud patterns are, therefore, only to be spotted from an orbiting satellite, and are constantly evolving and metamorphosing. For the geophysical scientist, they evoke an uncanny resemblance to star-shaped dunes in the desert.

26. ***Seascape*** **by Frederic Edwin Church, 1859**
A bank of rather lumpy *Stratocumulus* lying out over the sea; its slightly heaped texture distinguishes it from *Stratus*, while the setting Sun highlights (from below) some of the lumpiness on its underside. There is also a hint of *Stratocumulus lenticularis* lying just above the main cloud deck.

MARINE *STRATOCUMULUS* CLOUD BRIGHTENING

So far we've portrayed *Stratocumulus* as rather unexciting and even gloomy, but the good news is that it does have some redeeming properties, most notably, in its ability to help regulate Earth's climate. Lying at a low level, *Stratocumulus* is a relatively "warm" cloud, usually having a thickness of between 600 and 1,200 feet (200 to 400 m). It is therefore rich in water droplets, which significantly attenuate the transmission of sunlight through the cloud, and therefore also any warming effect, reflecting back to space a considerable proportion of incoming solar rays.

In addition, *Stratocumulus* being a low-level and "warm" cloud, its top surface radiates quite strongly in the infrared spectrum, causing net cooling. In these days of Arctic heatwaves, melting glaciers, and calving icebergs, *Stratocumulus* therefore plays a special role in helping to keep Earth cool.

A solution to the climate crisis?

Due to climate change, and because of its ability to cool Earth, there has been renewed interest in *Stratocumulus* in recent years, not least by those promoting climate intervention techniques

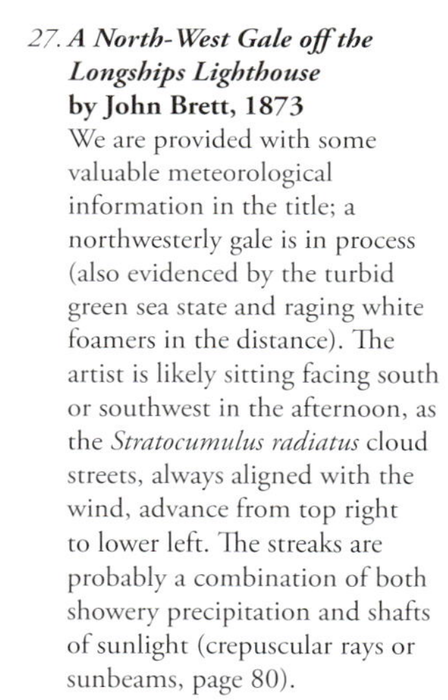

27. ***A North-West Gale off the Longships Lighthouse* by John Brett, 1873**
We are provided with some valuable meteorological information in the title; a northwesterly gale is in process (also evidenced by the turbid green sea state and raging white foamers in the distance). The artist is likely sitting facing south or southwest in the afternoon, as the *Stratocumulus radiatus* cloud streets, always aligned with the wind, advance from top right to lower left. The streaks are probably a combination of both showery precipitation and shafts of sunlight (crepuscular rays or sunbeams, page 80).

(as known as "geoengineering") to delay, halt, or even reverse global warming. This attention is warranted because it is quite easy, from a technical and engineering sense, to brighten marine *Stratocumulus* clouds and make them more reflective of sunlight through the Twomey Effect mechanism (page 73).

It has been hypothesized that the Twomey Effect could be achieved by a fleet of 1,500 boats continuously spraying a fine mist of seawater droplets, of about one-tenth of a millimeter in diameter, into the clouds. When implemented on a large enough scale, this would have the potential to increase Earth's albedo, and thus reduce global warming. Interestingly, the Twomey Effect can already be observed today on satellite images of Earth, having arisen inadvertently due to pollutants from cargo ships crisscrossing the oceans; these are regularly observed to leave bright trails in *Stratocumulus*, and may even cause a transition from open-cell to closed-cell *Stratocumulus*.

However, there is a major caveat, and not least through the law of unintended consequences: Climate modeling simulations indicate there might be some serious side effects and repercussions in other parts of Earth's climate system if marine cloud brightening were adopted as a climate intervention technique. The most concerning of these would be potential changes in the precipitation of the monsoons, on which billions of lives on Earth depend.

28.

NIMBOSTRATUS (Ns)

28. ***Foggy Winter Day. To the Left a Yellow House. Deep Snow* by Laurits Andersen, 1910**
A dull, gray, monotonous, overcast sky hangs low over this midwinter scene. A few streaks of falling precipitation (*virga/praecipitatio*) are discernible toward the center and right of the sky. These indicate that the cloud is most likely to be *Nimbostratus*, rather than (the usually non-precipitating) *Stratus*. The snow looks like it is beginning to thaw—perhaps a warm front is advancing across the region, with which *Nimbostratus* is commonly associated.

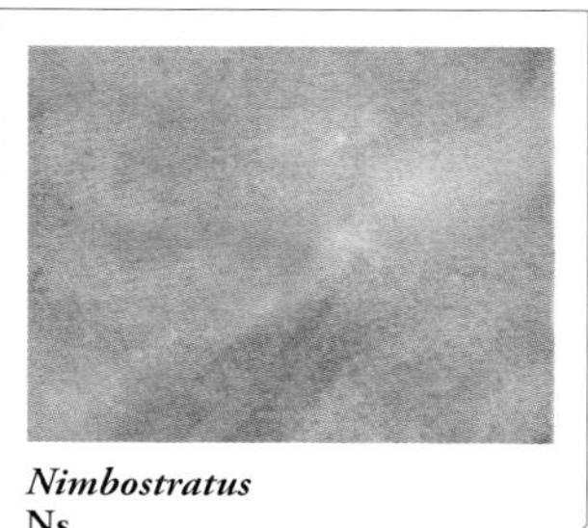

Nimbostratus
Ns

INDEX	
Genus	*Nimbostratus*
WMO codes	Not coded for C_L, except for its scud (C_L=7); otherwise C_M=2
Latin	"raining layer"
Species and varieties	None
Supplementary clouds	*praecipitatio* *virga*
Accessory clouds	*pannus*
Appearance	Very low, dull, wet
Frequency	Common

THE NATURE AND GEOMETRY OF *NIMBOSTRATUS*

Nimbostratus is a flat sheet of low, gray, featureless cloud extending right across the sky, which, at first sight, looks almost identical to *Stratus*. However, *Nimbostratus* has one defining characteristic that differentiates it from *Stratus*—it produces persistent precipitation; hence the *nimbo-* prefix (meaning "rain"). Typically, a steady, light-to-moderate rain or snow falls that may last for a few hours or more, often being associated with a weather front moving across the broad geographic region.

A considerable depth of cloud is usually required for the formation of precipitation in the atmosphere. *Nimbostratus* is therefore thicker than *Stratus*, and usually spans both the low and midlevels of the troposphere; in active weather fronts, it may reach into the high cloud level as well. This means less light from the Sun penetrates through the cloud to Earth, augmenting the dullness and dreariness experienced at ground level during daytime. Due to the falling precipitation keeping the air moist, the base of *Nimbostratus* is very low, usually less than 2,000 feet (600 m), and near the coast it may lie close to, or even at, sea level.

Being a largely featureless, overcast, gray layer, *Nimbostratus* has only one accessory cloud: *pannus*. It forms when the precipitation falling from *Nimbostratus* encounters a slightly warmer or drier layer of air close to Earth's surface, allowing the falling raindrops to evaporate slightly in the lowest few hundred feet (one to two hundred or so meters) above the ground. This evaporation causes cooling and therefore resaturation of the air in the immediate vicinity of the evaporating drops, leading to small ragged or "shredded" wisps of cloud appearing below the general cloud base. These continuously form and re-form, but remain entirely separate from the uniform deck of cloud lying just above it. *Pannus* may form wherever the evaporation of precipitation occurs, which is why it is also an accessory cloud of *Cumulus*, *Cumulonimbus*, and *Altostratus*.

The supplementary cloud features *praecipitatio* and *virga* can also be found with *Nimbostratus*. These specify whether the precipitation is reaching ground level, in which case *praecipitatio* is used, or if it has evaporated completely on descent, leaving visible streaks of *virga*.

29.

29. ***Before the Storm* by Thorsten Waenerberg, 1872**
Nimbostratus is most commonly found in weather fronts as a widespread, thick, uniform layer of precipitating cloud, covering the entire sky. Occasionally, however, large areas of stratiform precipitation may develop from *Nimbostratus cumulonimbogenitus*—that is, *Nimbostratus* formed by its progenitor, or mother cloud, *Cumulonimbus* (possibly depicted here in the far background). The lumpy clouds ahead of the advancing storm are likely to be *Stratocumulus* (possibly also *pannus*), of which both *Nimbostratus* and *Cumulonimbus* may also be a progenitor.

30.

CUMULONIMBUS (Cb)

30. ***Christmas Morning* by John Brett, 1866**
Ignoring the turmoil and the dark, low, ragged clouds (*Stratocumulus*) of the foreground, in the distance we can see the anvil tops of three isolated *Cumulonimbus incus* towers, tinged beautifully in the early morning sunlight. Crepuscular rays (page 80), cast by other cloud towers beyond the visible horizon, are depicted upon the right-most anvil. Winter *Cumulonimbus* are common visitors to windward coasts of Britain in cyclonic airstreams.

Cumulonimbus
Cb

INDEX	
Genus	*Cumulonimbus*
WMO codes	C_L=3, C_L=9
Latin	"raining heap"
Species	*calvus* *capillatus*
Varieties	*None*
Supplementary clouds	*praecipitatio* *virga* *incus* *mamma* *arcus* *murus* *cauda* *tuba*
Accessory clouds	*pannus* *pileus* *velum* *flumen*
Appearance	Threatening thundercloud
Frequency	Occasional in certain climates, infrequent or rare elsewhere

THE NATURE OF *CUMULONIMBUS*

The greatest and most potent of all clouds, without any doubt, is the mighty *Cumulonimbus* (from the Latin *cumulus* and *nimbus*, meaning "raining heap"), which can grow to heights of 50,000 feet (17 km) or more in summer.

The King of Clouds

Cumulonimbus marks the culmination, and literal high point, of the process of rising thermals or convection, which may have begun only an hour or two beforehand, in the case of the extreme convection. The process is quite simple: The first rising thermals becomes visible as small *Cumulus fractus* clouds, later *Cumulus mediocris*, before building into more powerful *Cumulus congestus* towers and eventually *Cumulonimbus*, given a suitable unstable atmospheric profile.

The fully developed and mature *Cumulonimbus*, residing at the pinnacle of the cloud hierarchy, sports a large crown on top of its head, just like all royalty. Commonly referred to as an "anvil" due to its visual similarity to a blacksmith's anvil (but probably more easily identifiable today as a flat-top haircut), it is officially named as the supplementary feature *incus*. It is an inevitable consequence of convection being capped, or limited from further vertical development, due to the presence of the strong, quasi-permanent inversion at the tropopause, where the troposphere meets the stratosphere (page 26). Here, the cloud cannot grow any higher, so instead it spreads sideways, and the ice crystals that have formed due to the air temperature now being below -40°F (-40°C; page 68) are blown downwind by the strong winds that are often present at these levels. This occurs because ice crystals live longer—and therefore blow further—than liquid water cloud droplets, as they sublimate more slowly than warmer droplets evaporate.

During winter in the middle latitudes and polar regions, the tropopause usually exists at a low altitude (4–6 miles/7–10 km), much lower than it does in warmer subtropical or tropical climates (7.5–11 miles/12–18 km). As a result, the absolute height of *Cumulonimbus* anvils varies considerably from region to region and from season to season. In general, the warmer the weather, the taller the *Cumulonimbus*, and the more severe the weather will be. Given the right conditions, a single *Cumulonimbus* can rapidly grow into a huge, powerful storm system covering thousands of square miles, bringing severe lightning, torrential rain, large hailstones, violent and gusty winds, and, more rarely, tornadoes.

31.

32.

33.

34.

Cloud Study **by Knud Baade, 1852**

31. Baade successfully depicts an atmosphere of raw, primeval emotion and melodrama in this almost realistic juxtaposition of both towering and decaying *Cumulonimbus*, set against the backdrop of crystal-clean polar sky—common features of Scandinavian or northern European coastal winter climates.

32. At left, *virga* (precipitation) falls heavily in a strongly sheared environment (the wind speed increasingly rapid with height, moving from left to right).

33. Above, the high-level *Cumulonimbus* anvil (feature *incus*) catches a few rays of low-lying Sun.

34. Right, a fresh *Cumulonimbus calvus*, its base lying below the horizon, appears to be glaciating into a *Cb. capillatus*. Patches of *Altocumulus* and *Stratocumulus* litter the remainder of the sky, detritus from earlier showers; they therefore can adopt the mother name *cumulonimbogenitus* (page 192).

THE GEOMETRY OF *CUMULONIMBUS*

The genus *Cumulonimbus* (often abbreviated to *Cb* by meteorologists) has only two species, *calvus* and *capillatus*, and no varieties. However, being the veritable granddaddy of all clouds, it has plenty of offspring in the form of eight supplementary features and four accessory clouds—the greatest numbers of any cloud.

The species *calvus* refers to a *Cumulonimbus* that has already attained its upper echelon status near the tropopause and has begun to flatten out, but whose top remains cumuliform in texture and does not yet appear to have frozen. One of the key visible aspects of such freshly formed *calvus* towers (as well as their antecedent *Cumulus mediocris* and *congestus*) is their very bright and highly reflective fractal upper surface, which look just like a cauliflower. These contrast strongly with their much darker cloud base, the disparity in light becoming increasingly marked as the clouds grow taller and deeper, and approach nearer to us. When this happens quickly, for example as a squall or thunderstorm suddenly advances, we may perceive it as particularly ominous or threatening. This is because it takes time for our eyes to adjust to rapid changes in light—usually about 15 minutes—the same duration, for instance, it takes for our eyes to adjust to the night sky after leaving the interior of a brightly lit home.

In common with deep, well-developed *Cumulus*, a *Cumulonimbus* may sport the accessory clouds *pileus* and *velum* (page 96), through which the *Cumulonimbus* continues to grow stridently and independently, behaving as though it were as a mountain top and pushing the ambient airflow around to its side (producing an extensive skirt of *velum*) or over its top (forming a hood of *pileus*). Meanwhile, *pannus* is another frequently spotted accessory of *Cumulonimbus*, being common to all precipitating clouds.

In contrast to *calvus*, *Cumulonimbus capillatus* always boasts a frozen anvil crown, which is easily identifiable by its somewhat fuzzier, fibrous, hairy form, typical of other glaciated clouds such as *Cirrus* (page 170). When the frozen head of a mature *capillatus* assumes this anvil shape (supplementary feature *incus*; page 128), a prominent and frequently observed additional supplementary feature on the lower surface of the overhanging anvil is breast-like or udder-like protuberances, termed *mamma* (from the Latin "breast"). To the uninitiated, these can appear as awe-inspiring as they are threatening, but usually the worst of the severe weather has already passed by the time they arrive. Their striking appearance is caused by lobes of sinking air that have become denser due to evaporation and cooling, leading to a reversal of the normal upward convection process.

35. ***Clouds* by Thomas Cole, 1838**
A powerful *Cumulus congestus* here grows into (likely already or soon) *Cumulonimbus calvus* (non-anviled/non-glaciated Cb). The atmosphere is unstable, with rain showers likely.

36. ***Study of Clouds with a Sunset near Rome* by Simon Alexandre Clément Denis, 1786**
Great towers of convection build explosively in the sky, yielding veritable cathedrals of *Cumulus congestus* and *Cumulonimbus calvus*. The air is rising so powerfully (from right to left, then vertically upward) that the towers overhang slightly on their windward edge. A powerful thunderstorm is imminent.

35.

36.

OTHER *CUMULONIMBUS* FEATURES AND ACCESSORIES

In the *International Cloud Atlas*, the height of a cloud's base determines the allocation of a low-, mid-, or high-level status. *Cumulonimbus* is therefore classified as a low-level cloud, despite it stretching upward into mid- and high levels as well. Occasionally, however, the convection that gives rise to *Cumulonimbus*-like storms can be generated entirely by dynamical processes in the mid-troposphere, independent of, and well above, surface influences such as daytime heating or evaporation from warm water bodies. For example, the release of energy at midlevels of the atmosphere may occur when a cool, dry layer overruns a warmer, moister layer, leading to the development of instability (or buoyancy) and powerful convection. This is what happens in the "dryline" events leading to severe thunderstorms over the Mid-West of the United States, or in "Spanish Plume" events in Europe.

In some very cold Arctic and Antarctic environments, a *Cumulonimbus capillatus* may be frozen entirely from top to bottom. This leads to the unusual appearance of a hairy, fibrous base (as well as an icy crown), as if a wall or curtain of snow is approaching along the ground. Visibility drops to nearly zero as the squall arrives, because frozen precipitation particles reduce horizontal visibility and obscure the view much more effectively than raindrops. In these polar winter climates, the tropopause often sits at the very low altitude of 4 miles (7 km), so the *Cumulonimbus* anvil forms much lower than in other climates.

As well as *incus* and *mamma*, other supplementary cloud features of *Cumulonimbus* include *praecipitatio*, *virga*, *arcus*, *tuba*, *murus*, and *cauda* (page 204), as well as the accessory cloud *flumen*. As we've learned already, *praecipitatio* simply refers to the fact that the cloud is producing precipitation (rain, hail, or snow); *virga* is used instead when the precipitation evaporates on descent and does not reach the ground. In contrast, both *arcus* (page 202) and *tuba* (page 206) are dramatic manifestations of a severe or violent *Cumulonimbus* storm system.

Like all active things in nature, in due course *Cumulonimbus* must die, but usually not before it has spawned several daughter convection cells in an iterative, self-regenerative process. Eventually, though, the supply of warm air and moisture feeding the cloud will be cut off, be that solar heating of the land surface by day, rising thermals over a warm water surface, or dynamical processes in the atmosphere. After some or many hours, all that remains are banks of stratiform cloud producing light to moderate precipitation, accompanied by low-level *pannus*, with the vestige of the once magnificent storm towers and anvils now reduced to just high-level streaks of *Cirrus spissatus* (page 170).

37. ***Cloud Study*** **by Knud Baade, date unknown**
Twin towers of *Cumulus congestus* rise into a region of pronounced windshear. There are also abundant rising wisps of *Cumulus fractus*, some of which may be in the foreground, closer to the artist, and therefore of lower altitude than the two towers. The clouds appear to be growing in an evaporative environment, drying out as they rise; if they were in a steady state, other clouds, such as *Cumulonimbus*, would surround them.

38. ***Port Ruysdael*** **by J.M.W. Turner, 1826**
It seems strange that a small sailing boat would set out in this weather! It must be at least a Force 5 or 6 on the Beaufort Scale—as evidenced by the choppy white waves. Visibility is heavily restricted—we cannot see the cloud base in the distance, the darkness of the horizon adding to the sensation that another squall is imminent. The tops of the cloud turrets indicate they are *Cumulus congestus* and *Cumulonimbus calvus* (center-top), although through the gaps there is a glimpse of patches of blue sky beyond, revealing abundant *Cirrus* or *Cirrostratus*—frontal weather systems are likely nearby.

37.

38.

39.

39. ***Marine* by Gustave Courbet, 1866**
This is an excellent capture of lengthy shafts of precipitation (*virga*) of various intensity descending from a likely *Cumulonimbus*. As the *virga* reach the ground, they are designated as the supplementary feature *praecipitatio*. Although the precipitation is heavy, the shower is brief; we can discern some brighter skies through the *virga* on the other side. The extreme rearmost edge of the convective (ascending) parts of the *Cumulonimbus* are visible at the top (center); this might be *mamma* too, or possibly the rear edge of *arcus*.

6

5

4

“The air up there in the clouds is very pure and fine, bracing and delicious. And why shouldn’t it be? it is the same the angels breathe.”

Mark Twain

3

2

1

MID-LEVEL CLOUD SPECIES

MID-LEVEL CLOUD FAMILY TREE

As we've learned already, the definitions of the mid- and high cloud levels (or *étages*) by the WMO are somewhat arbitrary. This is because the troposphere is very much deeper over the tropics compared to polar regions. All three cloud *étages* overlap somewhat and the concept of levels may be considered subjective. Here, we shall follow the WMO guidelines whereby mid-level clouds over polar regions lie in the range 6,500 to 13,000 feet (2,000–4,000 m), with the upper limit extended to 23,000 feet (7,000 m) in temperate regions and to 25,000 feet (8,000 m) in tropical regions. We will also follow the definition that the level is allocated by altitude of the base of the cloud, although this rule is not always adhered to in the literature.

Bearing this in mind, we find that the mid-level family tree comprises only two consistent mid-level genera, *Altostratus* and *Altocumulus*. This is the fewest of the three cloud levels—and a seeming meteorological poverty compared to the five genera at low levels and three at high levels! But paucity doesn't mean there is "less meteorology" happening at midlevels; again, it's only a result of the way that cloud levels are categorized—in reality, both of the low-level genera, *Nimbostratus* and *Cumulonimbus*, have more or less equal claims on midlevels as do *Altostratus* and *Altocumulus*, because they often extend through (and sometimes start) at the midlevel.

Altostratus and *Altocumulus*, while not being indicative of poor weather in the present, usually signal that a deterioration in the weather is on the way, particularly in the middle latitudes of Earth, where the passage of frontal weather systems is common. This is because wind speeds are generally greater the higher one goes up in the troposphere, so any forthcoming change in the weather is usually signaled initially at mid- and upper levels, ahead of changes at the surface.

Altostratus is a pale, diffuse layer of cloud, and, as hinted above, usually signals the imminent arrival of a weather front. It has no species but has five varieties (*translucidus*, *opacus*, *duplicatus*, *undulatus*, and *radiatus*), three supplementary features (*virga*, *praecipitatio*, *mamma*), and one accessory cloud (*pannus*).

CLASSIFICATION OF MID-LEVEL CLOUDS			
GENERA	SPECIES, VARIETY, MOTHER CLOUD, OR GENERAL OBSERVATION	*	o
Altostratus	*translucidus*	C_M=1	
Altostratus* & *Nimbostratus	(*Altostratus*) *opacus*	C_M=2	
Altocumulus	*translucidus* (not *duplicatus*) and predominating	C_M=2	
	perlucidus, and is continuously changing	C_M=4	
	Progressively invading the sky, thickening	C_M=5	
	cumulogenitus or *cumulonimbogenitus*	C_M=6	
	opacus or *duplicatus*	C_M=7	
	castellanus or *floccus*	C_M=8	
	of a "chaotic sky"	C_M=9	

* WMO code   o International cloud symbol

WMO codes, abbreviations, and respective symbols for selected mid-level cloud species. For example, if *Altocumulus castellanus* is observed, the code C_M=8 is recorded. Again, however, not all mid-level clouds are coded, and *Nimbostratus* is coded as mid-level. *Altocumulus* is also, perhaps, unduly over-represented.

1.

ALTOSTRATUS (As)

1. ***Plein Air-Painter at the Coast*** **by Robert Thegerström, 1881** Both *Altostratus* (smooth gray layers) and *Altocumulus* (dappled clouds) are most prominent in this depiction of a coastal scene in northern France by Thegerström. However, the bathers are using their sunshade, so one must assume that it is still a reasonably fine day. The pale upper background hue is suggestive of an advancing layer of thin *Cirrostratus*, so a deterioration in the weather may yet be forthcoming, but not for several hours.

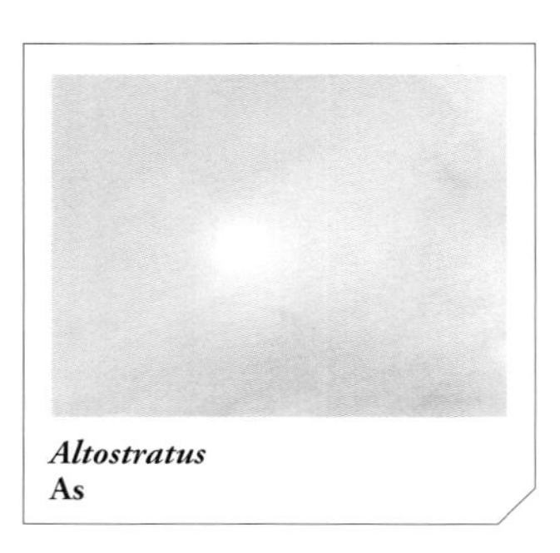
Altostratus
As

INDEX	
Genus	*Altostratus*
WMO codes	C_M=1, 2
Latin	"mid-level layer"
Species	None
Varieties	*translucidus* *opacus* *duplicatus* *undulatus* *radiatus*
Supplementary clouds	*virga* *praecipitatio* *mamma*
Accessory clouds	*pannus*
Appearance	Pale, diffuse layer
Frequency	Occasional

THE NATURE OF *ALTOSTRATUS*

As its name implies, *Altostratus* is a layer cloud with the apparent consistency of *Stratus* but found at midlevels of the troposphere. Much like *Stratus*, it appears as a diffuse and featureless layer that is pale white, light gray, or sometimes inky-bluish in color, with a fuzzy and indeterminable base. While *Stratus* exudes a dark and dreary tone, and often restricts horizontal visibility, *Altostratus* is usually high enough not to interfere directly with ambient conditions on the ground, other than dimming the sunshine. It gives the general feeling that a gradual deterioration of the weather is on the way, but not for several hours yet.

Altostratus may produce a light precipitation if associated with an approaching active weather front, but if it does, it is usually not inconvenient. More frequently, any precipitation evaporates as *virga* soon after descending from the base of the *Altostratus*, which is the reason for its fuzzy, diffuse aspect.

Varieties and supplementary features

Altostratus and its lower-level cousin *Nimbostratus* are the only cloud types that have no sub-species—a consequence of their ill-defined, indistinct, and nebulous appearances. Unlike *Nimbostratus*, however, which has no varieties and only two supplementary features, *Altostratus* boasts a total of five and two, respectively.

The first two of these varieties, *translucidus* and *opacus*, again refer to the transmission of light through the cloud. An advancing layer of *Altostratus* tends to weaken or blot out the direct rays of sunshine. This does not necessarily create an overall sensation of dullness, however, as mid-level clouds have a lower total water content than low-level clouds and therefore a significant proportion of incoming light still penetrates the cloud. Equally, if the *Altostratus* deck does not advance to cover the whole sky, the amount of diffuse light emanating from the rest of the sky tends to compensate for the reduction in direct solar rays. When this occurs, and if the Sun's disk is still discernible, the variety *translucidus* is appended to the cloud's name: if not, the variety *opacus* is used.

The other three varieties—*duplicatus*, *undulatus*, and *radiatus*—refer to their shape and form, as described in detail earlier (page 116).

The formation of *Altostratus*

One of the most frequent ways that *Altostratus* is formed is by gently rising air in cyclonic weather, such as along frontal boundaries and in low-pressure systems, features that we see regularly on weather maps. In these instances, the cloud forms as a result of gentle uplift, rising at a rate of only an inch or two (a few centimeters) per second or less—at least two orders of magnitude lower than the rapidly rising updrafts in a freshly formed *Cumulus congestus* or *Cumulonimbus*. The result is a smooth, laminar, fuzzy,

2.

and sometimes gently inclined layer of *Altostratus*, with its base gradually lowering as the weather system advances toward us.

Another common way for *Altostratus* to form is through "orographic forcing." This happens, for example, when an air mass arrives unimpeded from across the ocean and is suddenly forced to rise over a coastal mountain range or high ground. If the whole slab of the troposphere—its complete vertical profile comprising all three levels: low, mid-, and upper—are lifted simultaneously, the induced adiabatic cooling (page 48) may be enough to provoke saturation. If this occurs at midlevels, *Altostratus* is formed.

An alternative pathway to the formation of *Altostratus* arises due to the principle of hydrostatic balance (page 52). Being a layered deck, *Altostratus* sometimes gets left behind by one of its more boisterous cousins, such as a powerful thunderstorm, it being considered mere "meteorological detritus." For example, *Altostratus* is one of many clouds that may be produced by a mature *Cumulonimbus* but is then left abandoned long after the storm system has peaked and died away. When such cloud rebirth and metamorphosis occurs, *Cumulonimbus* is designated as the parent or "mother cloud" (page 192).

2. ***Coast Scene with White Cliffs and Boats on Shore* by J.M.W. Turner, undated**
In this rather hazy and overcast scene, the weather is fair. The principal cloud species appears to be *Altostratus*, possibly multilayered (variety *duplicatus*) but with some breaks too. The smoke plume indicates light airs and the sea surface appears calm, indicative of high pressure.

THE GEOMETRY OF *ALTOSTRATUS*

Altostratus has five varieties: *translucidus*, *opacus*, *duplicatus*, *undulatus*, and *radiatus*. We've discussed the first two varieties already in relation to the Low Clouds (page 108), although the method of formation of the latter three at midlevels, as well as their resulting appearance, differs considerably from those at low levels.

For *Altostratus*, *duplicatus* describes a variety where two or more separate layers coexist above (or below) one another—a situation that is common enough in stable and laminar airflows, though hard to spot from the ground if one layer hides the other. In general, at midlevels the variety *duplicatus* is best seen not with *Altostratus*, but when occurring in association with the much more magnificent and photogenic cloud *Altocumulus lenticularis* (page 156).

Radiatus is used with *Altostratus* to describe multiple laminar bands of the cloud, roughly aligned parallel with the wind: these may be formed orographically. Due to the generally higher wind speeds at midlevels of the troposphere compared with lower down, the cloud appears in streaks that may stretch completely across the sky. Due to perspective, the streaks appear to converge at a distance beyond the horizon. Although they all owe the origin of their alignment to features in the direction and flow of the wind, *Altostratus radiatus* does not look like either *Cumulus radiatus* or *Stratocumulus radiatus*, as it is a laminar and much more tenuous mid-level cloud than either *Cumulus* or *Stratocumulus*.

In contrast to *radiatus*, *undulatus* describes regular wave-like patterns forming perpendicular to the wind, having a short wavelength and which may form as billows (overturning waves on the top of a cloud layer)—this is the same as for low-level clouds. Again, however, the *undulatus* variety of *Altostratus* is much better expressed in the species *Altocumulus*, especially in the formation colloquially known as "mackerel sky" (page 154).

Both *Altostratus undulatus* and *Altocumulus undulatus* are sometimes confused with mountain wave clouds (*lenticularis*, page 156). However, the former can usually be identified by being transient and moving with the wind, rather than being geostationary. They also have a much shorter wavelength, typically of less than a mile (1.6 km) when compared to *lenticularis*, which have wavelengths of many miles. The special species *lenticularis* also does not occur with *Altostratus*; instead, it is found only at midlevels with *Altocumulus*.

There are just three supplementary clouds and one accessory cloud associated with *Altostratus*: *virga*, *praecipitatio*, *mamma*, and *pannus*, respectively.

3. ***Lake Siljan. Study*** **by Gustaf Wilhelm Palm, date unknown**
It is likely late in the day on a fine summer's evening, Palm capturing almost perfectly a serene atmosphere of calmness and tranquility, coupled with excellent visibility. A few layers of stratiform cloud, including patches of *Altostratus*, appear to be advancing gradually across the scene from upper left. The background sky color is an undeniable cyan, almost turquoise—typical of pristine Nordic environments.

3.

4.

ALTOCUMULUS (Ac)

4. ***Cloud Study at Sunset* by Frederic Edwin Church, 1873** Church excels at low-light productions, especially in the recreation of the characteristic aquamarine-cyan evening hues of a clean, dry polar continental airmass, a feature that he consistently repeats with near perfection. In this case, we are treated to a small patch toward the zenith. The numerous cloud decks include *Altocumulus undulatus* (billows; gray cloud in middle left). Above that, we find *Altostratus* or *Cirrostratus*; also *Cumulus mediocris* building into *Stratocumulus stratiformis* (center).

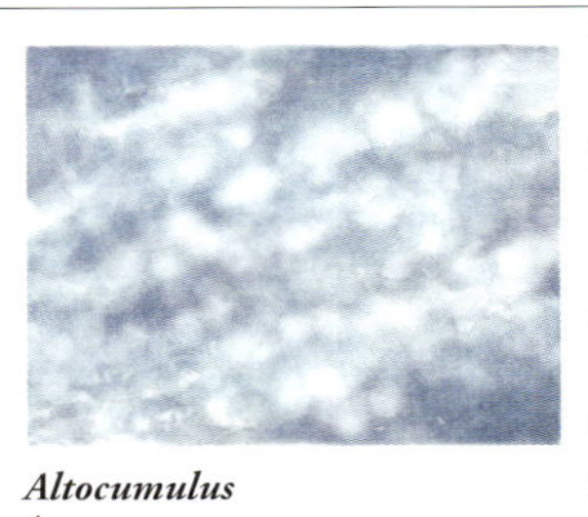

Altocumulus
Ac

INDEX	
Genus	*Altocumulus*
WMO codes	C_M=4, 5, 6, 7, 8, 9
Latin	"mid-level heap"
Species	*stratiformis* *lenticularis* *castellanus* *floccus* *volutus*
Varieties	*translucidus* *perlucidus* *opacus* *duplicatus* *undulatus* *virga* *radiatus* *lacunosus*
Supplementary clouds	*mamma* *cavum* *fluctus* *asperitas*
Accessory clouds	None
Appearance	Fine, dappled, broken layer
Frequency	Occasional

THE NATURE OF *ALTOCUMULUS*

Although the prefix *alto* means "high" in Latin, if you are a choral singer, you will know that alto singers do not possess the highest voice, instead they lie below the soprano, but above both the bass and tenor. In much the same way, *Altocumulus* can be thought of as lying above all the water-rich, low-level clouds, but not quite reaching the ice-pitched extremes of high-level cirriform cloud. Being one of the most common mid-level clouds, *Altocumulus* has many unique and distinctive mid-level cloud characteristics.

Altocumulus is a common and easy cloud to identify. It typically consists of a mid-level deck of small-to-moderate, slightly puffed-up cells of cloud, with small gaps between the cells that let the light through—this is the *perlucidus* variety).

How *Altocumulus* self-modifies

At first sight, *Altocumulus* may seem to drift along somewhat aimlessly on steady currents of wind in the mid-troposphere, without much change in its outward appearance. However, timelapse imagery reveals clues to its evolution.

Altocumulus is occasionally formed directly as an orographic cloud, in the same way that *Altostratus* is sometimes created—when an entire "slab" of air is lifted after passing over a hill or mountain range. The mountains do not have to be very high, just high enough to provide sufficient lift—and therefore adiabatic cooling—to saturate the air at one or more levels within the mid-troposphere, within which a cloud of very fine water droplets usually forms in less than a few microseconds. These cloud droplets grow quickly in size as they blow downwind. The very act of their condensation releases latent heat, which provides the impetus for a little convection to develop, leading to the formation of small convective cells. As air temperatures at mid-tropospheric levels are much cooler than those at the surface, but rarely below 4°F (-20°C), *Altocumulus* usually consists of supercooled water droplets.

5. ***Regatta at Argenteuil* by Claude Monet, ca. 1872**
Bulging yacht sails and specular reflection from the water surface (glint or scattered light, see page 80) tell us it is breezy day. The sky appears reasonably bright, however, with no sign of any threatening weather nor precipitating clouds. The single deck appears to be *Altocumulus translucidus*.

5.

Altocumulus also often forms during the gentle uplift of air, independent of any influence of surface topography. This typically happens in the presence of weather fronts, whereby relatively warm, moist, low-density air that approaches cooler, drier, denser air tends to ride up on top of it, rather than mixing with it, but at a very gentle, sometimes barely perceptible, rate of uplift. Once formed orographically, *Altocumulus* will usually continue to flow downstream with the wind (with the exception of the *lenticularis* species, which is geostationary).

Sometimes *Altocumulus* will form from a modification, or mutation, of *Altostratus*. This happens when *Altostratus* becomes increasingly subject to the laws of radiation—its base absorbing infrared radiation from below, while its upper surface emits it to space. These processes generate a slight convection (in much the same way as happens in marine *Stratocumulus*) and promote the development of small regular cloudlets, interspersed by small clear spaces. After about 20–30 minutes, the cloud may have completely morphed into what is termed *Altocumulus altostratomutatus* (we add *-mutatus* after the "mother cloud" name to describe this modification—see page 192 for more details on mother clouds).

WHY *ALTOCUMULUS* IS DIFFERENT

The key aspect of formation of *Altocumulus* is that, although it is characterized by convection—hence the *-cumulus* suffix—the convective currents within *Altocumulus* are much weaker and more horizontally regular than those found in *Cumulus congestus* or *Cumulonimbus*, which are more varied and often stronger by one or two orders of magnitude. This is because the formation of *Altocumulus* is not directly related to surface-based convection; the clouds are not caused by the direct heating of Earth's surface by the Sun, nor by convection from a relatively warm water surface such as the low-level clouds *Cumulus* and *Stratocumulus*.

THE GEOMETRY OF *ALTOCUMULUS*

Altocumulus has five species, seven varieties, and five supplementary features, making it second only to *Stratocumulus* as a cloud genus in terms of its range and total number of formations.

Like *Stratocumulus*, its five species are: *stratiformis*, *lenticularis*, *castellanus*, *floccus*, and *volutus*. *Stratiformis* is the most common species, describing a fairly uniform deck of gray–white, dappled, and slightly heaped layer of cloud composed of regular-sized cloudlets or cells that may or may not have gaps between them (varieties *perlucidus* and *opacus*, respectively). Unlike *Stratocumulus*, however, and because of its greater height above Earth, lower total water content, and usually thinner vertical extent, both sunlight and moonlight are more likely to penetrate *Altocumulus* to a greater degree (variety *translucidus*). *Altocumulus* is therefore not generally considered to be a "poor weather" cloud, although it may be a precursor of a deterioration in the weather.

As is best demonstrated within the closed-cell version of *Stratocumulus stratiformis*, the gentle and nuanced convection that sustains a layer of *Altocumulus stratiformis* is not initiated from below (as in *Cumulus*) but instead may be instigated from above by infrared radiation to space from the cloud tops. This causes the cloud to overturn in regular dappled cells or cloudlets, a pattern that is also repeated in *Cirrocumulus* (page 180). A layer of *Altostratus*, or even *Altocumulus lenticularis* mountain wave clouds (page 156)—if they last long enough—may morph into *Altocumulus stratiformis* through the same process.

The *Altocumulus* species *castellanus* (page 152) and *lenticularis* (page 156) are rather special; indeed, although both species are shared with *Stratocumulus* and *Cirrocumulus*, they are best exemplified with *Altocumulus*. Why? Because, being at midlevels, they are higher up than their low-level species and we are therefore more likely to see them, but not so high that they lack contrast when backdropped against even higher clouds. Also they usually manifest themselves in groups or clusters surrounded by attractive patches of blue sky, which is not guaranteed at the low level of *Stratocumulus*. Also associated with *castellanus* is *floccus*, which is also best displayed with both *Altocumulus* and *Cirrocumulus*. Sometimes referred to colloquially as a "parachute cloud," it describes cumuliform wisps or tufts of cloud (similar to *castellanus*) that also sport trails of precipitation (*virga*) extending downward from their bases, the whole cloud giving the impression of a parachute gliding effortlessly through the air. *Volutus*, meanwhile, is a rare, new cloud species (page 198).

As well as *perlucidus*, *opacus*, and *translucidus*, the other four varieties of *Altocumulus* are *duplicatus*, *undulatus*, *radiatus*, and *lacunosus*. The variety *undulatus* is quite common with *Altocumulus*: this describes stripes or billows in the cloud that are aligned

6.

perpendicular to the direction of the airflow, a formation more commonly known as "mackerel sky" (page 154).

Meanwhile, the five supplementary cloud features associated with *Altocumulus* are *virga*, *mamma*, *cavum*, *fluctus*, and *asperitas*. We have met *virga* (page 51) and *mamma* (page 130) already, while the three remaining features—*cavum*, *fluctus*, and *asperitas*—are all new clouds (pages 198–99).

Altocumulus, consisting largely of supercooled water droplets, is the ideal "exemplar cloud" in which to witness inadvertent cloud seeding because *cavum* (page 199), also known as "holepunch" or "fallstreak" clouds, when they occur, are most commonly seen in supercooled decks of *Altocumulus*. The small, regular-sized water droplets within *Altocumulus* also frequently give rise to coronae around the Sun or Moon.

6. ***Cloud Study, Sunset***
by John Constable, 1821
The principal gray clouds in focus appear to be patches of *Altocumulus*, the gentle convection within them released by mid-level instability. Cirriform clouds predominate in the background at upper levels, lit by the setting Sun, with possible *Stratus* or *Stratocumulus* featured in the lower quarter.

ALTOCUMULUS CASTELLANUS

One mid-level species of *Altocumulus* deserves our special attention, not just for its beauty but also because it is a well-known precursor, or prognosticator. The arrival of *Altocumulus castellanus*—referred to as "ice cream castles in the air" by Joni Mitchell in her famous 1969 album *Clouds*—gives us a clue as to how the weather of midlevels is changing. Winds are stronger at midlevels than at Earth's surface, so changes here often precede those at low levels. The appearance of *Altocumulus castellanus* in the sky therefore provides a fairly reliable weather forecast for the observer on the ground over the coming 12–24 hours.

Once known as *castellatus*, *Altocumulus castellanus*—the hint being in the name—consists of small but abundant, distinctive, medieval castle-style turrets, ascending upwards in the style of seashell whorls, or are perhaps more attractively described as upright mini ice cream cones. Their characteristic shape, with their height being greater than their width and their tops wider than their bases, is largely due to two physical processes, namely instability and expansion.

Provenance of *Altocumulus castellanus*

Instability arises at mid-tropospheric levels when a relatively cool layer of air spreads over and on top of a warmer, moister layer, destabilizing the boundary between the two air masses. The warmer air, being less dense, rises—then, when condensation occurs, it receives an added boost from the latent heat released at the moment of condensation.

In much the same way as bubbles of carbon dioxide rise valiantly and try to escape from a glass of pop just after it has been poured, each turret of destabilized mid-level air is forced to rise, trying to escape to a higher level at which it is no longer unstable. In doing so, however, it encounters decreasing air pressure and its natural response is to expand. This increasing expansion with height gives rise to its characteristic ice cream cone shape.

What we can learn from *Altocumulus castellanus*

In middle latitudes, *Altocumulus castellanus* often forms toward the end of a spell of warm, sultry weather. The visual evidence of increasing instability at midlevels that it provides is not just a forewarning—it is also a "nowcast" (a short-term weather forecast for 0–6 hours ahead) that cooler conditions have already arrived at midlevels.

If the advancing weather systems continue their onward procession—usually from west to east—the whole atmospheric profile, including that at low levels, may further destabilize. Thunderstorms and heavy precipitation are likely to break out, revealed in the greatest of all clouds, *Cumulonimbus* (page 126).

7.

7. ***The Dort Packet-Boat from Rotterdam Becalmed* by J.M.W. Turner, 1818**
The wind may be light and calm, and the weather probably warm, but the first intimation of a deterioration in conditions over the next 24 hours is visible in the sky above, provided by the well-known prognosticator, *Altocumulus castellanus* (top center). In the far distance, the crown of a possible *Cumulonimbus calvus* is almost discernible (center right).

Undulatus
Un

MACKEREL SKY: *ALTOCUMULUS UNDULATUS*

Due to its regular striped pattern, similar to that found on the back of the North Atlantic fish, *Altocumulus undulatus* is known commonly as "mackerel sky." These transient undulations, moving with the cloud, are arranged roughly parallel with one another, and can be separate or merged. They are caused by a rapid increase in wind speed with height, for example when a fast-moving current of air flows over another more languid layer. In such situations, and in much the same way as ocean waves are generated on the surface of the sea, oscillations develop on the interface between the two layers on the upper surface of the cloud. Differences in humidity and air density between the two disparate layers may also play a part. The strong wind on the upper surface then helps the cloud to overturn into regular cells, which are separated by a relatively short wavelength. When the white crests of each cell are lit up brightly by the Sun in an attractive fashion, they are known as "billows."

Occasionally, the waves curl up completely and overturn rapidly, just like an ocean wave does when it breaks on the seashore. They then become known as "Kelvin–Helmholtz waves" (after Lord Kelvin and Hermann von Helmholtz, who both studied turbulence). Their official supplementary cloud feature name is *fluctus* (page 198).

It is worth noting that *Altocumulus undulatus* can sometimes be mistaken for *Altocumulus lenticularis*. However, in the latter case the wave is geostationary (it stays fixed in the same place, relative to the ground) and the resulting wavelength is much longer than in *undulatus* or billow clouds, whose waves are transient and move with the cloud.

8. ***Sunset* by Samuel Palmer, 1861**
The scene is one of rural tranquility, but the background reveals a splendid portrayal of *Altocumulus undulatus* (mackerel sky).

9. ***The Beach at Villerville* by Eugène Boudin, 1864**
With everyone well wrapped up (or are they wearing the latest fashions?), it looks chilly for an excursion on the beach. However, the weather is fine and dry, and the skyscape is spectacular: The principal cloud deck is a broken layer of dappled *Altocumulus stratiformis*, tinged lightly by the setting Sun. It is not a mackerel sky; instead "salmon patchwork" might be a more appropriate and descriptive expression.

Is there truth in the proverb?

"Mackerel sky, mackerel sky, not long wet, not long dry"

If you have heard the prognostic proverb or some derivative of it, you may have wondered if there is any truth in it. In reality, proverbs and folklore tend to endure only if they are of some use! Long before the advent of modern weather forecasting, our forebears relied heavily on local knowledge being passed down through the generations, so that they might improve their livelihoods or simply increase their chances of survival. Today, a meteorologist might apply the term "nowcast" to a mackerel sky, as it provides evidence that both moisture and winds are increasing at midlevels of the troposphere, which usually heralds low-level changes in a few hours. But equally, as the wind is increasing, the unsettled weather ahead may pass by fairly quickly.

Mackerel sky can also be spotted in the high-level cloud *Cirrocumulus undulatus*.

8.

9.

Lenticularis
len

LENTICULAR CLOUDS: *ALTOCUMULUS LENTICULARIS*

One of the most unique and distinctive cloud species is *Altocumulus lenticularis*. Its characteristic flying saucer, or smooth lens shape can only be spotted over, or to lee of hills and mountain ranges, providing the perfect backdrop for photographers and artists seeking the ultimate image. On rare occasions, especially when tinted golden pink at dawn or dusk, its appearance is so other-worldly that they have been mistaken as alien UFOs.

Lenticularis is special because it forms only in the crests of the (otherwise invisible) atmospheric waves that are set up as air impinges upon, and traverses, mountainous terrain. Particular meteorological conditions need to be met for their formation, including pronounced atmospheric stability (either a rise in temperature with height or a reduced decrease, which curtails upward movements of air), increasing wind speed with height, and the necessary humidity for cloud condensation. Despite these constraints, satellite images of Earth confirm that *lenticularis* is ubiquitous, appearing in many parts of Earth on an almost daily basis.

10.

As already mentioned, the mountain waves that cause *lenticularis* are geostationary: that is, they remain anchored over the same location, in much the same way that the position of rapids and standing waves on the surface of a fast-flowing river remain stationary relative to an observer positioned on its bank.

In the crest of a wave

Lenticularis develops in the wave crests because the rising air cools adiabatically as it ascends into each crest; if the airflow contains enough moisture, saturation and condensation will occur rapidly (within a few microseconds) as the air temperature reaches a local minimum at the highest point in the crest. However, as soon as the air exits the crest and starts to descend into the neighboring trough, the cloud quickly evaporates as the air warms adiabatically once again on its descent. This regular "up and down" oscillatory process, or resonance, then propagates forward with the airflow, more or less undisturbed. This means that, although the lenticular clouds themselves are geostationary, the air continuously flows through them.

Depending on the speed of the airflow through the wave, the time it takes for air to traverse a single wavelength from crest to crest, also known as its frequency, typically varies between 5 and 20 minutes. The amplitude, or vertical displacement, of each wave depends on both the topography and the meteorological attributes of the airflow itself, but deviations of 1,650 feet (500 m) between trough and crest are not uncommon. The steepness of the wave also depends on factors such as the topography, wind speed, and height of the temperature inversion; the first few waves are generally the steepest. Cliff-like, near instantaneous vertical displacements (known as "hydraulic jumps," page 188) are known to occur in highly turbulent airflows.

Why so smooth?

The laminar and smooth-edged appearance of *lenticularis* can be attributed to the stable atmospheric conditions necessary for their formation and the small, fine droplets of which they are composed. As the air is forced through each wave crest, the rapid condensation results in small cloud droplets, which quickly evaporate again after only a minute or two, or even less, as the air exits the crest and dives into the following trough. Their short lifetime means they have little opportunity to grow into bigger droplets nor have much of an influence on their nearby environment—they evaporate just as quickly as they condense, meaning the cloud stays smooth and streamlined in appearance.

The species *lenticularis* is found at all three cloud levels, so its parent genus may be *Cirrocumulus*, *Altocumulus*, or *Stratocumulus*.

10. ***Mount Discovery, with open leads in the new ice*** **by Edward Wilson, 1911**
Mount Discovery lies southwest of McMurdo Sound in Antarctica. Here we are looking directly upwind through six or seven oscillations of trapped lee waves (*Altocumulus lenticularis*, assuming their bases lie above 6,500 ft/2,000 m), which have been created by wind blowing over the mountain. This means the gap (or wavelength) between each wave crest is about 6 miles (10 km), typical of mid-level lee waves.

TRAPPED LEE WAVES

Due to its close association with atmospheric stability, *lenticularis* is a fair-weather cloud. Its appearance in the sky may be used as a predictor, at least over the short term, for warm and dry conditions, although it may be windy too.

Due to their greater total water content, the low- and mid-level versions *Stratocumulus lenticularis* and *Altocumulus lenticularis* appear bright on their upper surfaces, but tend to have patches of gray on their underside: especially *Stratocumulus lenticularis*, which has a dark base. *Cirrocumulus lenticularis*, on the other hand, is entirely white.

When regular trains of *lenticularis* appear downwind of mountain ranges, they are known by atmospheric scientists as "trapped lee waves"—they are "trapped" because the energy of each wave remains largely confined within the stable layer of air in which they form, the energy dissipating only slowly downstream. Indeed, it is not uncommon for their wave trains to continue unbroken for more than 10 or 15 wavelengths downwind, thus covering well over 100 miles (160 km). On occasion, lee waves have been observed on satellite images stretching more than 600 miles (1,000 km) downstream of their initial disturbance.

Variation in lee waves

Lee waves that are set off by single isolated peaks—such as a small island or a prominent isolated volcano—leave a "wake" in the clouds, just like a small boat leaving a wake on a calm water surface: a pattern also readily identified by satellite. Alternatively, when impinging upon the edge or corner of a mountain range, the cloud tends to develop more of a herring-bone pattern downstream, with the orientation of the waves becoming angled into the flow, reminiscent of the way that light waves "diffract" around a corner or through a narrow slit. Wave interference may lead to additional and more complicated patterns of cloud downwind.

In contrast, when the impinging airflow rises uniformly over a long mountain ridge aligned perpendicular to the wind, the resulting *lenticularis* may appear as a narrow but long and straight roll of cloud, repeating downstream: a formation that can sometimes be confused with *radiatus*.

As the physics of lee waves dictates, their wavelength (the crest-to-crest distance between adjacent *lenticularis*) increases with higher wind speeds, and because wind speed generally increases with height above Earth, so the wavelength of *lenticularis* also generally increases with altitude. Typical values range from approximately 1–5 miles (1.5–8 km) at low levels, 4–12 miles (7–20 km) at midlevels, and up to 20 miles (32 km) at high levels.

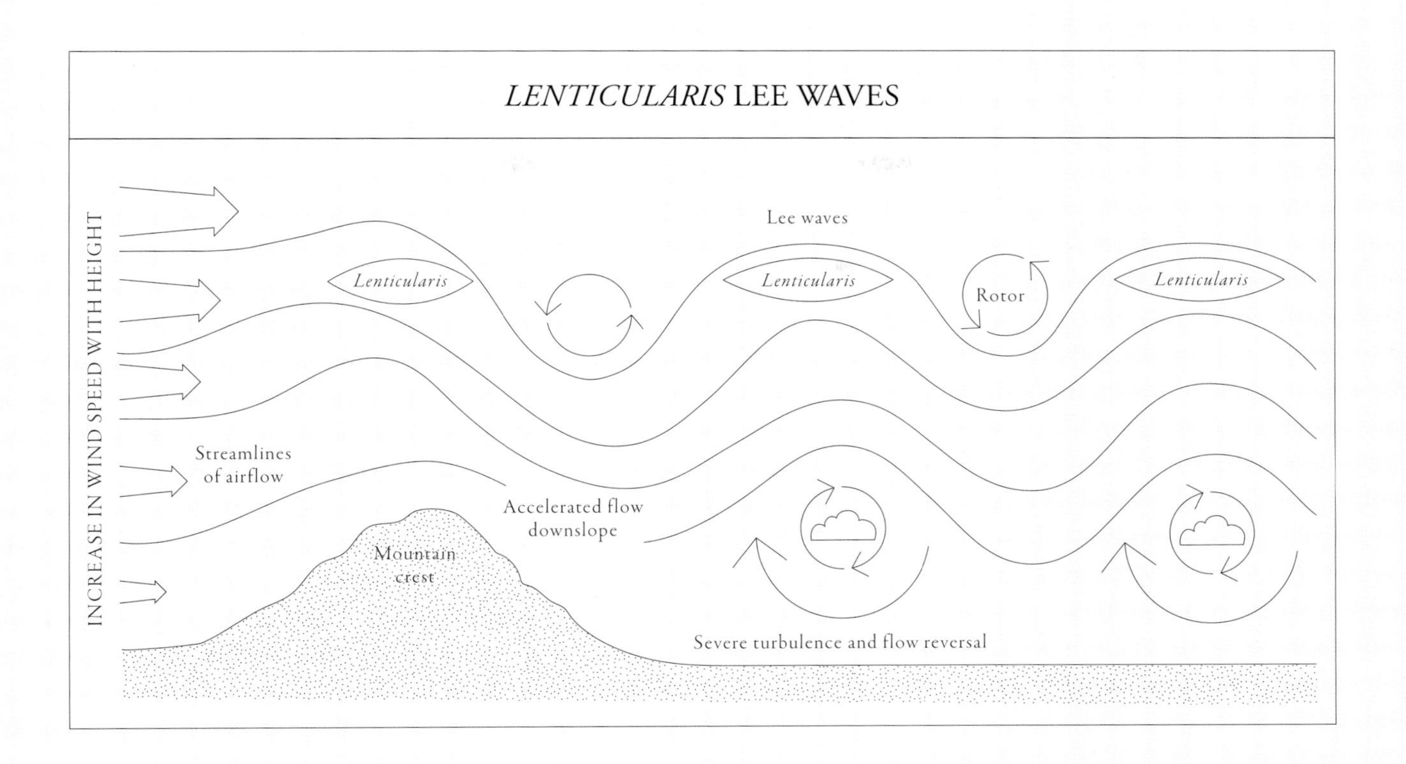

Schematic of an airflow traversing a mountain range and the formation of lenticularis due to adiabatic cooling in its lee wave crests
Mountain waves may form downwind of any hill or mountain when there is increased stability with height, combined with increasing windspeeds. Each lee wave is geostationary; it remains in the same place as the air blows through the cloud. In extreme cases, an additional "rotor" cloud may form within the orbital circulation directly below each wave crest, leading to local flow reversal and severe turbulence at ground level.

When *lenticularis* breaks the rules

Although *lenticularis* is almost always geostationary, there are a few occasions when the cloud may change position. The first case arises during strong winds when a prominent wave crest, and its associated lenticular cloud, is positioned a short distance downwind of a rounded mountain summit, and the wave crest begins to act as if it were the mountain peak itself. When this occurs, the air will continue to flow into the wave, but the wave crest (and its associated wave train propagating downwind) will, ever so slowly, begin to migrate downstream. Eventually, a point will arrive when the airflow cannot continue rising into the first wave because it has moved too far downstream, and its axis therefore is too heavily tilted with respect to the mountain summit below it. When this happens, the system collapses, only for a fresh primary wave crest to form back over, or just to the lee, of the mountain summit. The resulting wave train will also shift back upstream a short distance, before the exercise repeats itself.

ALTOCUMULUS LENTICULARIS DUPLICATUS: *PILE D'ASSIETTES*

One of the most beautiful, alluring, and eye-catching displays of any cloud must surely be the *duplicatus* variety of *lenticularis*, which is more commonly described using the French term *pile d'assiettes* (its literal translation is somewhat less endearing: "a pile of plates"). It describes stacked, or duplicated, layers of lenticular clouds that look as if they are piled up on top of one another.

As *lenticularis* forms in a stable air mass and is a direct result of the airflow being forced over a hill or mountain, the hydrostatic balance of the atmosphere (page 52) ensures that after passing over the brow of the mountain, it immediately sinks back down the other side (hydrostatic balance actually overdoes it; in its rather impetuous desire to restore order immediately, the air, rather than returning directly to rest after crossing the mountain, continues to oscillate for some considerable distance downstream, forming lee waves). In such instances, the various micro layers of the atmosphere flowing over the hill or mountain tend not to mix with one another, but instead are all lifted discretely at the same time, as if in unison. When only the topmost portions of each micro layer saturate, then a series of individual lenticular clouds will form, the crest of each one apparently stacked upon its nearest neighbor below. The result is the cloud formation *pile d'assiettes* or *Altocumulus lenticularis duplicatus*. *Pile d'assiettes* can also occur at low and high cloud levels.

It does not require particularly high mountains for *lenticularis* to form. For example, in coastal or humid environments when air is forced to rise over quite modest hills of only several hundred feet in height, there is usually sufficient lift to saturate the air at many levels. If the airflow is stable enough and combined with adequate wind shear—increase of wind speed with height—mountain wave clouds usually form within a few microseconds.

As *lenticularis* clouds consist of tiny water droplets or ice crystals that have no opportunity to grow any larger before they evaporate again, they sometimes display striking iridescent colors (page 78) that are best exemplified in stratospheric mother-of-pearl, or nacreous, clouds (page 208).

Occasionally, long-wavelength but small amplitude *Altocumulus lenticularis*, which live for a little longer in atmosphere, may visibly display the effects of the laws of radiation. The cooling of their upper surface by infrared radiation to space causes small overturning cells or mini cloudlets to become manifest in the cloud layer (*Altocumulus perlucidus*). Equally, strong winds blowing across the cloud's upper surface may roll the cells into wavy "billows" (*Altocumulus undulatus*), which themselves may break, forming the beautiful Kelvin–Helmholtz wave and new supplementary cloud feature *fluctus* (page 198).

11. ***Lenticularis duplicatus Cumulostratus* by Luke Howard, date unknown**
Howard brilliantly captures the exact form of *pile d'assiettes* (*Altocumulus lenticularis duplicatus*) in this schematic sketch. Below the wave clouds are some low-level *Stratus* and *Cumulus*; at high levels are some depictions of cirriform clouds.

11.

6

5

"I am the daughter
of Earth and Water,
And the nursling of the sky;
I pass through the pores
of the ocean and shores;
I change, but I cannot die"

Percy Bysshe Shelley, "The Cloud" (1820)

4

3

2

1

HIGH CLOUD SPECIES

HIGH-LEVEL CLOUD FAMILY TREE

There are three high-level cloud genera: *Cirrus*, *Cirrocumulus*, and *Cirrostratus*. Together with the anvil top of the towering *Cumulonimbus*, they represent and highest altitude that our weather-producing clouds can attain. This is because they are restricted from further vertical growth by the presence of the tropopause: a strong temperature inversion at the boundary between the troposphere and overlying stratosphere. This creates an effective "lid" or "cap" of enhanced stability, preventing their further vertical growth.

The WMO define the high cloud levels as dependent on latitude: 10,000–25,000 feet (3,000–8,000 m) for polar regions; 16,500–45,000 feet (5,000–13,000 m) in temperate climates, and 20,000–60,000 feet (6,000–18,000 m) for tropical regions. If you glance quickly back at the low- and mid-level definitions (pages 86 and 138), you'll see immediately that the mid-level limits overlap with those of the high level. Also, from the high-level cloud definitions above, it can be seen that the tropical troposphere can be more than twice as thick at the equator than it is at the poles!

The first of the high-level cloud family is the genus *Cirrus* (meaning "a lock of hair"). It has five species (*fibratus*, *uncinus*, *spissatus*, *castellanus*, and *floccus*), four varieties (*intortus*, *radiatus*, *vertebratus*, *duplicatus*), and two supplementary features (*mamma*, *fluctus*).

Cirrocumulus (meaning "a heaped lock of hair") has four species (*stratiformis*, *lenticularis*, *castellanus*, and *floccus*), two varieties (*undulatus*, *lacunosus*), and three supplementary features (*virga*, *mamma*, *cavum*).

Last but not least is *Cirrostratus* (meaning "a hairy or fibrous layer"). It has only two species (*fibratus* and *nebulosus*), two varieties (*duplicatus*, *undulatus*), and no supplementary features.

Interestingly, *Cirrus* and *Cirrostratus* are always completely frozen and therefore consist solely of ice crystals, though *Cirrocumulus* may occasionally consist of supercooled water droplets (especially the species *castellanus* and *lenticularis*), but it will usually freeze after a couple of hours.

CLASSIFICATION OF HIGH-LEVEL CLOUDS

GENERA	SPECIES, VARIETY, MOTHER CLOUD, OR GENERAL OBSERVATION	*	o
***Cirrus* (Ci)**	*fibratus* or *uncinus*	$C_H=1$	
	spissatus (non-*cumulonimbogenitus*), *castellanus* or *floccus*	$C_H=2$	
	spissatus cumulonimbogenitus	$C_H=3$	
	uncinus or *fibratus* progressively invading the sky	$C_H=4$	
***Cirrostratus* (Cs)**	Progressively invading the sky but extends <45° above horizon	$C_H=5$	
	Progressively invading the sky, extends >45° above horizon, but not whole sky	$C_H=6$	
	Covering the whole sky	$C_H=7$	
	Not progressively invading the sky, not covering whole sky	$C_H=8$	
***Cirrocumulus* (Cc)**	Alone in the sky, or predominating	$C_H=9$	

* WMO code   o International cloud symbol

WMO codes, abbreviations, and respective symbols for selected high-level cloud species. For example, if *Cirrus uncinus* is observed, the code $C_H=1$ is recorded. However, not all high-level clouds are coded.

1.

CIRRIFORM CLOUDS

1. ***Morning in Spring with north-east Wind, at Vevey* by John Ruskin, 1849 or 1869**
The high-level striated clouds (*Cirrus fibratus radiatus* with a hint of *Cirrocumulus*, thickening to *Cirrostratus* and perhaps even some *Altostratus* or *Altocumulus*, upper left) are likely forming in a strong north or northwesterly high-level jetstream.

CIRRIFORM CLOUDS

The higher you go in the troposphere, the colder it becomes. Despite the presence of abundant supercooled liquid water droplets with a temperature below freezing point (32°F/0°C) in clouds at both low and midlevels of the troposphere, by the time the threshold of -36°F (-38°C) is reached, all hydrometeors are found in a frozen state. And given that the air temperature near the tropopause is typically somewhere between -40°F (-40°C) and -85°F (-65°C), condensed water vapor, at least at upper regions of the troposphere, is nearly always found in a frozen state, typically in the form of tiny hexagonal ice crystal columns or platelets. Exceptions do occur, such as in supercooled decks of *Cirrocumulus*, or when liquid water droplets are exported into upper levels of the troposphere by powerful convection currents in *Cumulonimbus*, but given the opportunity, they will freeze into an icy mixture within a few minutes.

Snowflakes and ice crystals fall at gentle speeds and more slowly than raindrops: something that is observed during a flurry of snow. The terminal velocity of a snowflake at Earth's surface is not much more than 3 feet (1 meter) per second, compared with 25–35 feet (8–10 meters) per second (up to 25 mph) for the largest raindrops. At high levels of the troposphere, the ice crystals are much smaller and are colder than at ground level, have less mass, and are more varied in size and shape; here, the terminal velocities are even lower, reaching a maximum of only perhaps 1½ feet (0.5 meter) per second for 1 mm-sized crystals in the upper troposphere.

The molecular bonds holding ice crystals together are also stronger than those in liquid water droplets. This, together with their slower fall speed, means that ice crystals last longer in the air than raindrops, and therefore have the potential to cause a greater cooling effect upon their nearby environment as they fall.

Although the air may be saturated at high levels—the evidence being visible as a cloud—due to the low air temperature at such heights, the total absolute amount of water present remains low because cold air cannot hold much water vapor (pages 26 and 48). This means that ice clouds are optically thin, usually remaining white or translucent, and tenuous in nature. However, when clouds consist solely of ice crystals, they have a large range of habits and facets, and may yield some spectacular optical effects and haloes (page 182).

As a result of the morphology of ice crystals and their behavior at low temperatures, the high-level tropospheric clouds of *Cirrus*, *Cirrocumulus*, and *Cirrostratus*—commonly referred to as the "cirriform clouds"—are often quite different in shape, form, aspect, and duration to mid- and low-level clouds that consist entirely of water droplets. At the same time, and especially against a backdrop of a deep cobalt-blue sky, they add to the enormous variety, diversity, and beauty of the skies.

2. ***Scene on the Loire, near the Coteaux de Mauves* by J.M.W. Turner, ca. 1830**
The Sun's disk is weakened but still discernible through an extensive layer of white or pale cloud; this is therefore *Cirrostratus* or *Cirrocumulus*. There is a hint of a sun pillar (page 183) directly above the Sun, indicating a cloud composed of ice crystals. Over the water surface, the atmosphere is misty or hazy with restricted horizontal visibility, as is so often the case in Constable paintings. Specular reflection of sunlight off the water, as well as the lack of sail, indicate almost calm conditions.

3. ***Yacht Racing in the Solent* by Alice Maude Taite Fanner, 1912**
Thick streaks of icy white *Cirrus fibratus radiatus* stretch across a fine and fresh blue sky. At top left, there appears to be some *vertebratus* (or possibly *undulatus*), its components arranged perpendicular to the upper airflow direction, from lower left to upper right. A few puffs of daytime *Cumulus* occupy a much lower level (center and right of sky). It is windy both at the surface and at high-cloud levels; the *Cirrus* likely being steered by strong jetstream winds.

2.

3.

Cirrus
Ci

INDEX	
Genus	*Cirrus*
WMO codes	C_H=1,2,3,4
Latin	"lock of hair"
Species	*fibratus* *uncinus* *spissatus* *castellanus* *floccus*
Varieties	*intortus* *radiatus* *vertebratus* *duplicatus*
Supplementary clouds	*mamma* *fluctus*
Accessory clouds	None
Appearance	White, wispy, feathery
Frequency	Common

THE NATURE AND GEOMETRY OF *CIRRUS*

Meaning "lock of hair," *Cirrus* aptly describes the soft, fleecy, fibrous appearance of these high-level ice clouds. Due to the long lifetime of ice crystals in the atmosphere, as well as the strong winds that often blow at upper levels of the troposphere, a large assortment of cloud shapes and forms are produced, meaning that there are five different species of *Cirrus*, the joint highest of any cloud genus: *fibratus*, *uncinus*, *spissatus*, *castellanus*, and *floccus*.

The *fibratus* species describes straight or slightly curved white icy streaks or filaments. The filaments or strands are usually separate from one another, and often occur in association with the variety *radiatus* (see below). In contrast, *uncinus* describes white fibrous, elongated striations that stretch or fall downward from a pronounced hook or tuft, the whole cloud element sometimes resembling the shape of an extended comma punctuation mark, colloquially known as "mares' tails" (page 172). Both of these *Cirrus* species, *fibratus* and *uncinus*, are regularly observed at high levels.

Spissatus is the only species of cirriform cloud that is thick enough to appear grayish—this is because it is usually an offspring of the anvil of a *Cumulonimbus incus*. The storm cloud itself may have died away many hours beforehand, leaving behind a dense residue of ice crystals at upper levels, which over the intervening period may have been steered a considerable distance away by strong winds. Strong convection in active weather fronts may also produce *Cirrus spissatus*, becoming visible to observers on the ground when strong winds blow it well ahead of the weather system.

The remaining two species, *castellanus* and *floccus*, also appear with *Cirrus*, but are perhaps better exemplified with the separate genus of *Cirrocumulus*.

Varieties and features

There are four varieties of *Cirrus*—*intortus*, *vertebratus*, *radiatus*, and *duplicatus*—and two supplementary features—*mamma* and *fluctus*. *Radiatus*, *duplicatus*, and *fluctus* occur with no fewer than five different cloud genera, and *mamma* with six (see the Cloud Table on pages 14–15). The varieties *intortus* and *vertebratus*, however, are both unique to *Cirrus* and therefore deserve a special mention.

Intortus describes the confused or irregular arrangement of *Cirrus* filaments, or their apparent irregular arrangement when viewed obliquely from ground level. Turbulence and windshear (rapid changes of wind speed and direction with height), combined with ice crystals gently falling through the layers of windshear, are the likely causes of their perceived entanglement. Similarly, *vertebratus* describes the arrangement or apparent arrangement of *Cirrus* filaments that form the shape of a fish skeleton or a rib cage,

4.

which can arise as the direct result of real atmospheric processes acting at 90 degrees to one another (such as we have already met with *radiatus* and *undulatus*, respectively). Equally the formation may appear briefly in the sky, and like *intortus*, may arise simply as a result of perception or the viewing perspective of the observer on the ground.

Finally, the species *Cirrus radiatus* has the unique status of being the only cloud made in large quantities by humans—because it is most commonly manifest as airplane condensation trails, or linear "contrails," although it is not exclusively made this way. When it is human-made, we append the mother cloud name *homogenitus* to its nomenclature (page 194), its full title becoming *Cirrus radiatus homogenitus*.

4. ***Study of Cirrus Clouds* by John Constable, ca. 1822**
Arguably, the principal focus of here is not cirrus at all! Instead, it's the prominent white puffs that take precedence. These are, plausibly, *Altocumulus castellanus* or *Altocumulus floccus* given their lack of glaciation (*castellanus* and *floccus* do occur at cirriform level too, but are usually much smaller than depicted here). Somewhat secondary to the *castellanus* puffs, *Cirrus fibratus radiatus* dissects the painting from left to right, as if driven by a powerful westerly jetstream (which it often is). However, the direct juxtaposition of it with the species of *floccus* and *castellanus* appears meteorologically incongruous.

ETYMOLOGY

Google Ngram Viewer tells us that the term "mares' tails" has been in widespread use since at least 1800, and possibly even before 1600. The word "mare" originates from Old English "mere" or "mearh," itself of Celtic and Germanic origin. A drop in usage is apparent after 1900 and again after 1950, as automobiles finally replaced the horse as a trusted mode of everyday travel.

MARES' TAILS: *CIRRUS UNCINUS*

One of the most common *Cirrus* cloud formations is *Cirrus uncinus* (meaning "hooks" of *Cirrus*), or "mares' tails." These are elongated white feathery striations or filaments that stretch downwind from a pronounced hook or tuft. They are not indicative of imminent poor weather, as the fact that we can see the high-level cloud from below in the first place means that it must be a relatively fine day. Nevertheless, mares' tails tend to form during changeable conditions and periods of mobile weather systems. This is because their formation is linked with strong upper tropospheric winds, or the jet streams, which tend to guide our weather systems along upper-level "conveyor belts," at least in the middle latitudes.

The small white tufts of *Cirrus uncinus* represent the site or position where cloud formation has been initiated—where saturation and condensation of the air has taken place—usually in the form of ice crystal nucleation. They therefore represent locations of slight upward movements in the air that may be produced by convection (instability), for example when a layer of colder air flows over a warmer, moister layer. Equally, gentle upward motions may be caused by the near-lateral advance of a weather front (pages 34 and 108), or they may be the consequence of an earlier gentle uplift of the whole air mass over mountains. Whatever the cause, the result of these processes means that *Cirrus uncinus* tends to appear in linear clusters, or groups, across the sky, rather than in isolated formation. If the uplift and convection at *Cirrus* level is stronger and more pronounced, the separate species of both *Cirrus* and *Cirrocumulus*, namely *castellanus* and *floccus*, may be formed instead.

The tails of *Cirrus uncinus* are effectively trails of precipitation (*virga*, although not denoted as such), being tiny ice crystals that have encountered wind shear by falling into a region of stronger or weaker wind, or wind from a different direction, and within which they are swept along laterally. Given the generally longer lifetime of ice crystals in the atmosphere compared with water droplets, these tails can therefore stretch considerable distances downwind before they evaporate (or sublimate) completely.

5. ***Cirrus in Different Forms* by Edward Kennion after Luke Howard studies, date unknown**
"Locks of hair" of *Cirrus uncinus*, as depicted in Howard's sketch of 1803. "Mare's tails" generally form in clusters, stretching and extending in approximately the same fashion and direction, all of which lends a certain asymmetry to each scene.

6. ***Cloud Drawing* by Luke Howard, date unknown**
In this instance, the "locks of hair" are somewhat randomly arranged with no preferred order or direction. If the cloud elements are of a "skeletal" impression or shape, then the variety *vertebratus* is appended to the cloud name. Alternatively, if they are completely irregular and distorted, the variety *intortus* is used.

5.

6.

7.

CIRROSTRATUS (Cs)

7. ***Cloud Study*** **by Frederic Edwin Church, 1880**
At upper center and left, we can see a few patches of *Altocumulus undulatus* (billows). Slightly lower down and to the right, the prominent pink line may indicate the position of the leading edge of a higher but incomplete *Cirrocumulus lenticularis*, backdropped by a general layer of *Cirrostratus*.

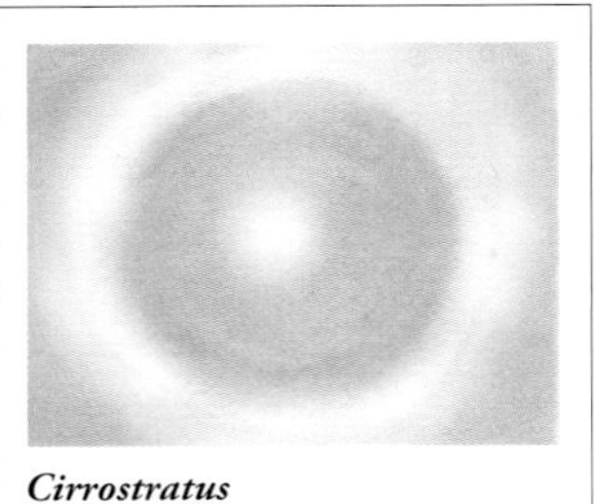

Cirrostratus
Cs

INDEX	
Genus	*Cirrostratus*
WMO codes	C_H=5,6,7,8
Latin	"layer of locks of hairs"
Species	*fibratus* *nebulosus*
Varieties	*duplicatus* *undulatus*
Accessory clouds	None
Appearance	Nebulous pale white diffuse layer
Frequency	Common

THE NATURE OF *CIRROSTRATUS*

In common with its lower-level cousins *Stratus* and *Altostratus*, a layer of *Cirrostratus* has a somewhat fuzzy and rather nebulous outward countenance. It usually appears as a diffuse veil of smooth or fibrous elements, which may gradually spread across the sky, providing the first visual clues of an advancing weather front that may still lie hundreds of miles away. Unlike its lower-level relatives, however, *Cirrostratus* is composed entirely of ice crystals, lending it a much more tenuous composition than the lower, water-rich clouds, always appearing white and translucent to the eye.

There are two species of *Cirrostratus—fibratus* and *nebulosus*. Just like *Cirrus fibratus* from which it often originates, *Cirrostratus fibratus* consists of thin filaments or striations, but in this case these are contained within the veil of cloud itself. In contrast, *Cirrus nebulosus* has no horizontal variation in its tone whatsoever. Indeed, it is sometimes so tenuous and indistinct that it is not readily observable unless the shadow of a higher cloud, such as a contrail, is cast upon it: or the cloud itself may only be revealed by the materialization of a beautiful ice halo (page 182).

There are also just two varieties of *Cirrostratus—undulatus* and *duplicatus*. Owing to the frequent presence of strong winds in the upper troposphere and sudden changes in wind speed and direction with height, the *undulatus* variety, consisting of regular, short-wavelength overturning cells or billows (page 154), is seen quite commonly in *Cirrostratus*, but more especially with *Cirrocumulus*. In addition, given the translucent nature of *Cirrostratus*, the *duplicatus* variety (consisting of two or more decks of the cloud) is also observed more easily than it is with lower cloud decks.

There are no supplementary clouds or accessory features associated with *Cirrostratus*.

8. ***Storm at Rügen* by Hans Gude, 1882**
A thin layer of *Cirrostratus nebulosus* is hardly distinguishable from the surrounding sky. It may only become noticeable when the Sun's rays or moonlight are dimmed and become more diffuse after shining through it. Here, a lower layer of cloud (perhaps *Stratus*, or *Altostratus*) also obscures the view to the horizon.

9. ***Cloud Study with Roof Tops* by John Constable, undated**
It is tempting to infer some cloud iridescence in this Constable, but it is more likely to be just characteristically early or rather late in the day, with the high-level *Cirrostratus* patches tinged a pinky cream by the oblique sunlight as a result. More interesting meteorologically is the prominent low-level gray cloud positioned above the rooftops; it looks like a poorly formed *Stratocumulus lenticularis*, perhaps in the process of "modification."

8.

9.

10.

CIRROCUMULUS (Cc)

10. ***Sunset Across the Hudson Valley, Winter*** **by Frederic Edwin Church, 1870**
Above the mountain (center), a large bank of fair-weather and high *Cirrocumulus stratiformis* drifts by (or perhaps *Altocumulus translucidus perlucidus*, depending on the exact level). Directly beneath, smaller and thinner gray patches of cloud, possibly *undulatus*, together with a few low-level scraggy patches of *Cumulus fractus* or *Stratocumulus*, lie above the mountain summit. The upper two cloud layers appear to start forming at a distinct "boundary" or "line" in the sky; maybe the remains of a weather system or orographically forced. In the far distance, a thickening veil of *Cirrostratus* approaches.

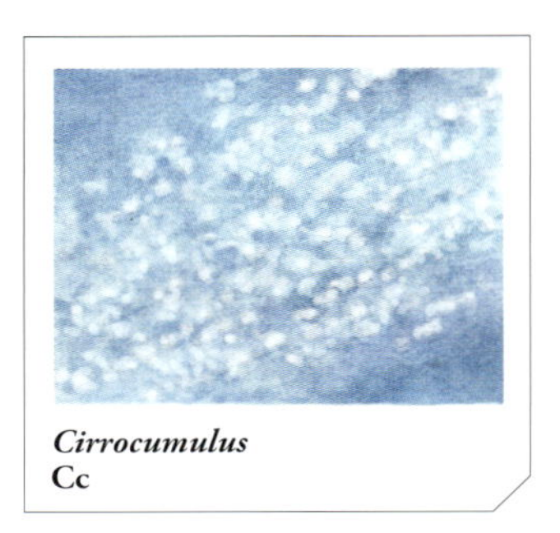
Cirrocumulus
Cc

INDEX	
Genus	*Cirrocumulus*
WMO code	C_H=9
Latin	"lock of hair, heaped"
Species	*stratiformis* *lenticularis* *castellanus* *floccus*
Varieties	*undulatus* *lacunosus*
Supplementary clouds	*virga* *mamma* *cavum*
Accessory clouds	None
Appearance	Dappled, high white layer of fine weather
Frequency	Occasional

THE NATURE OF *CIRROCUMULUS*

Broadly speaking, *Cirrocumulus* can be thought of as the high-level equivalent of *Altocumulus* (see pages 148–161). Due to similar formation mechanisms, all four species of *Cirrocumulus*—*stratiformis*, *lenticularis*, *castellanus*, and *floccus*—are shared with *Altocumulus* (and with *Stratocumulus*), in addition to its two varieties, *undulatus* and *lacunosus*. However, given the higher and colder level of formation of *Cirrocumulus*, its largely ice crystal composition means that there are some notable differences in both form and texture. Indeed, it might be said that these differences in composition and appearance allow it to become, arguably, the most beautiful of the cloud genera.

The genus *Cirrocumulus* describes a finely dappled layer of mini white cloudlets at high altitude. It never appears gray, nor does it darken the sky appreciably, and so its appearance continues to give the impression of a fine, or indeed glorious, day. The absorption and emission of infrared radiation, which forms a vital part of Earth's climate system, also plays an important role in the evolution of *Cirrocumulus*, as it frequently mutates from, or into, other cirriform clouds.

Its most common species, *Cirrocumulus stratiformis*, initially looks much like *Altocumulus stratiformis*, but it is thinner and always white, with its individual cloudlets appearing relatively small due to their great height and distance away from the ground. The cloudlets are usually separated by clear gaps (less than a mile across) that allow blue sky to penetrate from above. If the cloudlets are arranged in regular wavy stripes or billows, these also describe the variety *undulatus* (best exemplified in the "mackerel sky" formation, page 154). Increasing wind shear, which is common at high levels, helps to overturn the cloud in such circumstances. Meanwhile, the *lacunosus* variety, as was the case with the mid- and low-level clouds, describes a honeycomb structure within *Cirrocumulus*, there being more clear air than cloud.

Beautiful and alluring

As is the case with *Altocumulus lenticularis* (page 156), *Cirrocumulus lenticularis* also describes beautiful lens-shaped mountain wave clouds that frequently form at high cloud levels to the lee of a mountain ridge when atmospheric conditions permit; the waves may eventually propagate into the stratosphere too. If the cloud droplets or ice crystals are tiny—which is common enough with *lenticularis* and all high-level clouds—iridescence (cloud coloration due to the diffraction of light; page 78) may be present when viewed from a certain angle, adding further to the spectacle. *Cirrocumulus lenticularis* is also often associated with a unique and powerful feature that sometimes forms directly over mountains: the "hydraulic jump" (page 188).

11.

The two remaining species of *Cirrocumulus*—*castellanus* and *floccus*—are equally beautiful and alluring. They too form in a similar way and have an outward mien somewhat analogous to their mid-level relatives, *Altocumulus castellanus* and *Altocumulus floccus*. Unlike them, however, *Cirrocumulus castellanus* and *floccus* are always white, and appear more fleecy and tufty, lacking both the fractal and cauliflower structure of lower-level cumuliform clouds; when this is noticeable, it provides evidence of their ice crystal constitution.

Because of the vertical movements present within *Cirrocumulus*, and unlike both *Cirrus* and *Cirrostratus*, *Cirrocumulus* may display visual evidence of convection and thus precipitation may fall from it (although it never reaches the ground)—usually this originates within the *castellanus* or *floccus* species. *Mamma* may also be seen occasionally. If supercooled water droplets are present in a layer of *Cirrocumulus stratiformis*, given the opportunity, the droplets will readily freeze into ice crystals, falling out of the cloud and leaving a "holepunch," or *cavum* (page 199), behind in the cloud layer.

11. ***The Beach at Villerville, Normandy*** **by Axel Lindman, ca. 1878**
Occasionally some ambiguity may be allowed when attempting to distinguish between *Cirrocumulus* and *Altocumulus*, notwithstanding the fact that altitude limits of mid- and high cloud levels overlap (page 164). A good cloud-spotting tip worth remembering is that the small cloudlets of *Cirrocumulus* always remain white, whereas those of *Altocumulus* tend to be slightly duller. Here, in this Impressionist scene of women working on the beach at low tide, the cloud layer is therefore more likely to be *Cirrocumulus stratiformis*.

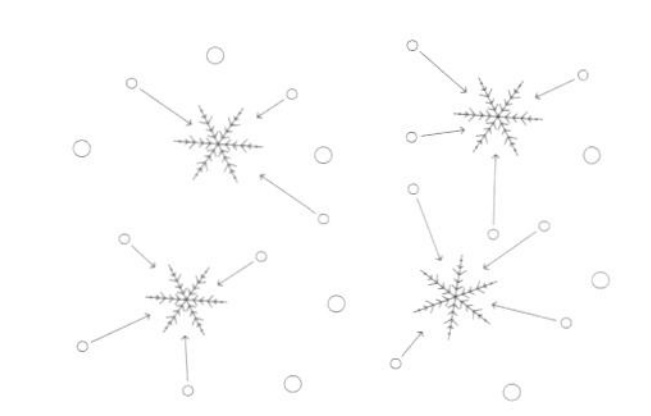

Ice crystals grow at expense of water droplets

Schematic depiction of ice crystals growing at the expense of supercooled water droplets, when the air is saturated with respect to ice but not with respect to liquid water.

ICE CRYSTAL HALOES

Ice crystal clouds, such as *Cirrostratus*, occasionally give rise to stunning optical illusions in the atmosphere. This is because ice crystals interact more with rays of light than water droplets.

Unlike water droplets, which in the atmosphere are round or oblate in shape, ice crystals have many different facets (or *habits*), with sharp angles and edges delineating the boundaries between the crystal faces; clouds comprising billions of tiny ice crystals are therefore effectively a veritable floating hall of minute mirrors, causing any incoming rays of light to potentially be reflected and/or refracted in a huge variety of ways. The result can be any one of a large range of beautiful optical effects, each one depending on the size, shape, and orientation of the ice crystals, the intensity and direction of the incoming light, and the viewing angle of the observer.

The most common ice crystal displays seen in cirriform clouds, at least when viewed from the perspective of the middle latitudes and tropics, are the "common" 22-degree solar or lunar halo, "sun dogs," "sun pillars," and the circumzenithal arc.

The common halo

The 22-degree halo, or common halo, is routinely seen as a whitish ring encircling the Sun or Moon at that precise angular separation away from the celestial object. It is caused by hexagonal ice crystals tumbling slowly downward with no particular orientation. Judged by sailors and seafarers over the centuries as a portent of deteriorating weather, the common halo is most obvious and best seen when a thin veil of *Cirrostratus nebulosus* advances across the sky, often in association with an approaching warm weather front.

Closely related to the common halo are sun dogs—their official name being *parhelia*, or *parhelion* (singular), from the Greek *para* meaning beside and *helios* meaning Sun. *Parhelia* are a pair of "mock suns" or pronounced bright spots in the sky, often colorful, that are found equidistant on either side of the Sun; they also occur at an angle of 22 degrees and thus intersect the common halo if it is present. *Parhelia* are again caused by horizontally aligned plate crystals, fluttering downward like millions of tiny slips of paper, or leaves in the fall, but keeping their horizontal alignment as they do so. When appearing as moon dogs on either side of the Moon, they are known as *paraselene* (from the Greek for "beside the moon").

WILSON AND THE BRITISH ANTARCTIC EXPEDITIONS

Edward Wilson (1872–1912) traveled twice to Antarctica with Captain Robert Falcon Scott, firstly as a junior surgeon and zoologist on the Discovery Expedition (1901–04) and later as Chief Medical and Chief Scientific Officer of the ill-fated Terra Nova Expedition (1910–13). He was also a talented artist. Wilson later died in his tent with his comrades, Captain Robert Falcon Scott and Henry Robertson Bowers, on their return from the South Pole in March 1912.

Sun pillars and circumzenithal arcs

Sun pillars are vertical beams of light, most frequently seen close in time to sunrise or sunset, that stretch upward as if emanating from the Sun itself, especially when it lies near to, or just below, the horizon. Again, they are caused by horizontally aligned ice crystals. Sun pillars can also be seen occasionally during heavy snow showers in a broken sky, looking toward the setting Sun, when large snowflakes also align themselves horizontally as they fall to Earth. When viewed from an airplane, a lower sun pillar may be seen to project vertically downward from the Sun.

Finally, the impressive circumzenithal arc is caused by both plate-shaped and horizontally aligned hexagonal crystals, but you will need to crane your neck upward in order to spot one, as it lies directly overhead in the zenith. If you are lucky, you will see a brightly colored arc of what is sometimes described as a "smile in the sky" or an "upside-down rainbow." As the name implies, these only occur in the zenith and only with a solar elevation of less than 32 degrees.

All of the above optical displays can be observed fairly frequently in cirriform cloud, especially in *Cirrostratus nebulosus*, due to the regular composition of its constituent ice crystals. Sun dogs, sun pillars, and the circumzenithal arc all require the presence of horizontally aligned plate crystals, so when one of them is present, it is worth scanning the sky to look for the other two at the same time. Both sun dogs and sun pillars are best observed when the Sun is low in the sky, ideally near sunrise or sunset.

12.

12. **Watercolor of *paraselene* (or "moon dogs") together with several other haloes, as seen at 9:30 p.m. on June 15, 1911, at Cape Evans (Antarctica) by Edward Wilson**
Here, the *paraselenae* connect with the common 22-degree halo on either side of the Moon. A 46-degree halo is also present. The upward-facing curve (connected to the top of the common halo) is part of an Upper Tangent Arc. A moon pillar (vertical pillar of light above and below the Moon) is also evident.

13. ***Hampstead Heath Looking Toward Harrow* by John Constable, 1821–22 (overleaf)**
As is common in Constable paintings, diffuse light dominates, the Sun being reduced to a pale disk shining weakly through a layer of high-level cirriform cloud—we know the cloud must be cirriform and is frozen because a Sun pillar is visible directly below the Sun. A few clumps of *Altocumulus* left of center also hint at a forthcoming deterioration in the weather.

14.

RARE ICE HALOES AS CELESTIAL PORTENTS

Some ice haloes are particularly rare; these, and their associated optical effects, are generally seen only in the cold seasons in polar, continental, or high-mountain climates. Due to the much lower temperatures in these regions, and therefore the greater incidence of frozen clouds at all levels, ice haloes are seen more frequently there and are often more complex. Ice clouds may even occur at ground level—for example, the stunning, sparkling, and ethereal phenomenon known as "diamond dust" is a relatively common winter phenomenon in Canada and parts of Scandinavia. It occurs in a thin, largely invisible, ice fog—each crystal glittering and glinting beautifully in the light. It often yields a wide range of stunning haloes and other surreal optical effects.

There are many rare ice haloes, each with the capacity to astound though a range of optical effects. These include "light pillars" (not the same as Sun pillars), the "sub-Sun," the parhelic circle, the 46-degree halo, tangent arcs, Parry arcs, and many others. Many of these are rare, typically being observed only once in a lifetime even by polar scientists and explorers.

Light pillars and the sub-Sun

Light pillars are the equivalent of Sun pillars (page 183) but occur only at night when artificial lighting in towns and villages appears to project colorful "laser beams" of light vertically upward. They need to be viewed from a distance, with a prerequisite being the presence of a thin ice fog consisting of horizontally aligned hexagonal or plate-shaped crystals. Rare trumpet-shaped light pillars may occur when column-shaped ice crystals align themselves along their horizontal axes, giving the very real and eerie impression of a searchlight being beamed downward from a spaceship above!

Another rare halo is the sub-Sun, although it is more commonly seen today thanks to modern air flight as it may be spotted from an airplane when looking downward upon a bank of *Cirrostratus*. It appears as a bright spot of light directly below the Sun at the same angle below the horizon as the Sun itself is above it, and is caused by the direct reflection of the Sun's rays off the upper surfaces of horizontally aligned plate crystals, acting in much the same way as if a large mirror (or billions of tiny ones) were placed in the same position.

The parhelic circle and Parry arcs

The parhelic circle is another unusual halo phenomenon, appearing as a line of white light stretching around the whole sky at the same altitude as the Sun (or, even more rarely, the Moon). It may be seen when the Sun is low in the sky and is caused by the reflection of

14. ***The Discovery*** **by Edward Wilson, published on** ***The Voyage of the Discovery*** **book cover, 1905**
A rare halo display silhouetted by *The Discovery*, on the First Antarctic Expedition (1901–04). Haloes visible include the common 22° halo, the 46° halo, part of the parhelic circle (horizontal line), sun dogs (*parhelia*), a rare upper sunvex Parry arc (above the 22° halo), and possibly two infralateral arcs (left and right of the 22° halo).

light off near-vertical faces of hexagonal ice crystals that intersect the *parhelia* ("sun dogs"—page 182) at the 22-degree common halo (page 182).

The Parry arcs were first documented in modern times by Sir William Edward Parry in 1820 during one of the many Arctic expeditions in search of the Northwest Passage. Parry arcs are rare and complex, changing shape and moving position relative to the Sun, depending on the solar altitude. Parry arcs require a dense cloud of ice crystals and often appear in close formation with what are called upper tangent arcs. Their presence requires column-shaped ice crystals to keep both their long axes and upper and lower prism faces remaining in the horizontal, meaning their degrees of freedom are restricted—they may rotate only around a vertical axis.

15.

Seeing the future in the sky

Throughout history, and in common with the sudden appearance of astronomical phenomena such as comets or supernovae, rare and beautiful celestial haloes in the heavens have been interpreted as portents or omens, heralding major societal changes, for better or for worse. According to popular belief, just before the Battle of the Milvian Bridge in 312 CE, Emperor Constantine is said to have looked up to the Sun and seen a cross of light emblazoned with the words (in Greek) "through this sign, [you shall] conquer." It is perhaps too easy for us now to explain this as some sort of spectacular solar halo display—maybe it was. Regardless, Constantine won a decisive victory that changed the course of European and world history, as the following year he officially recognized Christianity as a permissible religion in the Roman Empire.

Similarly, at the Battle of Athelstaneford in Scotland in 832 CE, King Angus faced off an army of Saxons. Looking up, he saw a white cross in the blue sky akin to the saltire or St. Andrew's Cross. The King prayed to St. Andrew, pledging that if he won the battle, he would make Andrew the patron saint of Scotland, or so the legend goes. Needless to say, the Scots won the battle, and so the saltire became the flag of Scotland.

15. ***Haloes*** **by Edward Wilson, Antarctic Expedition of 1910–13**
Frozen clouds are prerequisite for haloes. Outside of the polar regions, they are therefore usually seen only in high cirriform cloud layers. When the air is intensely cold and stable, however, ice crystals may remain in suspension undisturbed. When this happens, rare and spectacular haloes may be observed. In these sketches and watercolors, we can observe the common 22° halo, *parhelia*, sun/moon pillars (vertical lines), parts of the parhelic circle (horizontal line), upper and lower tangent arc ("gull's wing"), as well a possible rare Parry arc and infralateral arcs (top image).

Orographic cirrus

OROGRAPHIC CIRRUS AND THE HYDRAULIC JUMP

When a fast-flowing fluid such as water encounters an obstacle, the flow immediately over and downstream of the obstacle can become greatly disturbed and perturbed. This results in a sudden jump in the height of fluid, a reduced average flow velocity, and significantly increased turbulence downstream. We see this transition frequently in rivers, when a steady, laminar, quiescent current is transformed into a raging torrent immediately after passing over a hidden obstacle or following a rapid change in its flow regime. Physicists and fluid engineers describe the fluid flow at this point as "supercritical" and the sudden rise in the height of the fluid as a "hydraulic jump."

It is the same with the atmosphere, the fluid this time being air and the obstacle being a mountain range. Given strong enough wind shear with height and a suitably stable atmospheric temperature profile, a hydraulic jump may form directly over, or a bit upstream of, a mountain barrier. The mountain range does not need to be high—hydraulic jumps are often seen in satellite imagery forming over the small hills of England and Ireland. The most important factor is the vertical profile of temperature and wind speed, which, if suitable, allows the orographic waves to propagate vertically right up to the tropopause, and occasionally even beyond it into the stratosphere. A steadily increasing wind speed and stable atmospheric profile—that is, a reduced drop in temperature with height compared to normal—provides the best conditions for vertical wave propagation.

Hydraulic jumps at *Cirrus* level are relatively easily to identify from both ground level and from above when viewed from an airplane or seen on satellite images. This is because the sudden lifting in airflow accompanying the hydraulic jump, which is often many thousands of feet, leads to rapid adiabatic cooling and saturation of the air mass, resulting in unique cloud formations that can span all three cloud levels. The hydraulic jump also results in severe turbulence; before the phenomenon was fully understood by fluid dynamicists in the 1960s and 1970s, many airplanes and lives were lost in the turbulence associated with hydraulic jumps.

At the tropopause, the hydraulic jump is often marked by a sharp and pronounced line of *Cirrocumulus lenticularis*, more commonly known as "orographic cirrus" by atmospheric scientists. When viewed in timelapse animation, the cloud appears to stream away continuously from the mountain ridge, or even slightly upwind of it as the vertically propagating wave tilts slightly backward with height, the initial point of formation remaining geostationary. However, the cloud does not stream from the mountain peak: instead, it is formed near the tropopause at the crest of the jump in air, far above the highest mountain peaks.

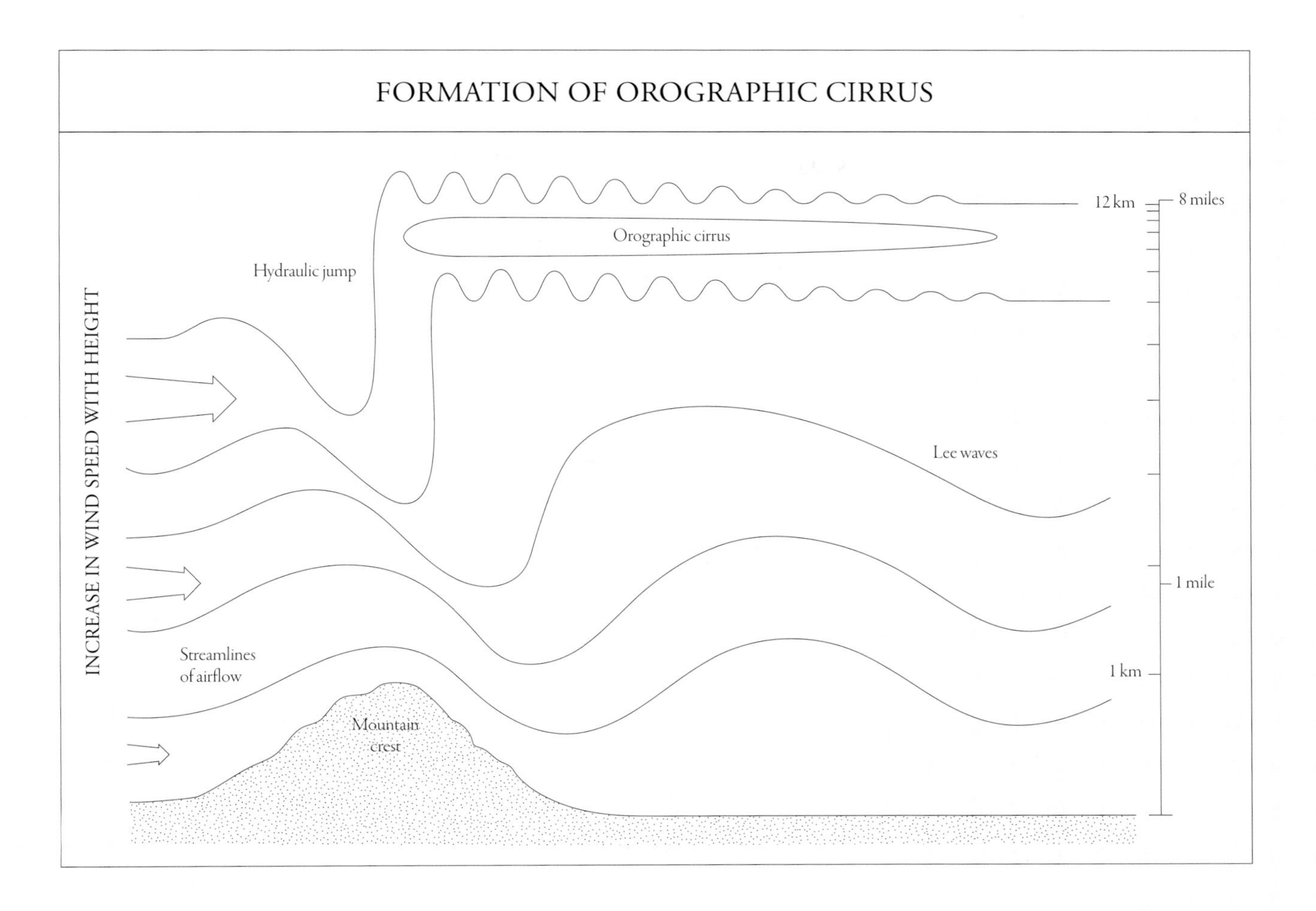

Along the eastern ridges of Rocky Mountains of North America, especially between Montana and Alberta, the cloud even has a special name: the Chinook Arch.

In contrast to some of the "trapped" mountain lee wave clouds, such as *Altocumulus lenticularis* (page 156), the orographic cirrus of a hydraulic jump is an "untrapped" mountain wave cloud. Its wave energy is not confined to a single layer and therefore it continues to propagate, dissipating much more quickly than it does with trapped waves, and so the long, regular wave trains associated with low- or mid-level lee waves generally do not occur with orographic cirrus. Instead, the cloud particles, particularly if they have frozen, tend to be swept downstream sometimes by 1,000 miles (1,600 km) or more, before slowly evaporating or sublimating away.

Schematic of the hydraulic jump and the formation of orographic cirrus
In certain circumstances (increasing wind speed with height; reduced fall of temperature with height), the airflow traversing a mountain barrier becomes "supercritical," leading to a sudden rise, or "hydraulic jump," in its level, together with increased turbulence, downstream. The sudden cliff-like rise in the airflow causes adiabatic cooling and the formation of orographic cirrus.

6

5

4

3

2

1

"Wherever the blue heaven is hung by clouds, or sown with stars, wherever are forms with transparent boundaries, wherever are outlets into celestial space, wherever is danger, and awe, and love, there is Beauty"

Ralph Waldo Emerson, "The Poet" (1884)

RARE AND UNIQUE CLOUDS

MOTHER CLOUDS

We all know people whose business cards and online profiles are furnished with letters after their name indicating university degrees, memberships, society fellowships, and so on. It is the same with some special clouds, known as "mother" clouds, whose names are adorned with various suffixes from which we are able to infer their provenance.

The WMO's *International Cloud Atlas* confers such special "mother" names on all ten of the principal cloud genera when it is possible to determine their origin. These clouds are then allocated a special "mother" cloud name and suffix when we are able to infer the cloud genus or process that it has been generated by (if so, append the earlier cloud genus name or process, followed by *-genitus*), or when the cloud is a new mutation, or complete modification from another cloud genus, or process (if so, add the previous cloud genus name, or process, followed by *-mutatus*).

Cloud evolution and mutation

Clouds are always evolving and mutating in the sky. Back in 1803, Luke Howard had already recognized the innate ability of clouds to morph from one type to another, referring to these "modifications" in his famous treatise.

To give an example, on a bright early morning in middle latitude climates when there is a moist and unstable lower troposphere, it is quite common for *Cumulus* clouds to grow rapidly and develop into *Stratocumulus* clouds, which then quickly spread out across the sky, rapidly blocking out the sunshine by mid-morning. This happens because the vertical development of rising thermals is checked by an inversion only a few thousand feet higher up, leading to their tops spreading out horizontally. These clouds would then be described as *Stratocumulus cumulogenitus*—i.e. *Stratocumulus* generated by *Cumulus*. If there was no inversion, however, and the thermals continued to rise to mid- or upper levels of the troposphere in a completely unstable environment, then the resulting large *Cumulonimbus* cloud might be described as a *Cumulonimbus cumulogenitus*.

The same is true for all ten cloud genera, although only certain clouds can morph into others, and usually it is a one-way journey with no turning back!

THE MOTHER CLOUD TABLE

Genera (type)	-genitus	-mutatus
Cirrus e.g. *Cirrus homogenitus* (*Cirrus* that has formed from an airplane contrail)	*Cirrocumulus* *Altocumulus* *Cumulonimbus* *Homo*	*Cirrostratus* *Homo*
Cirrocumulus e.g. *Cirrocumulus homomutatus* (A contrail that has transformed completely into *Cirrocumulus*)		*Cirrus* *Cirrostratus* *Altocumulus* *Homo*
Cirrostratus e.g. *Cirrostratus cumulonimbogenitus* (*Cirrostratus* that has formed from the anvil of a *Cumulonimbus*)	*Cirrocumulus* *Cumulonimbus*	*Cirrus* *Cirrocumulus* *Altostratus* *Homo*
Altocumulus e.g. *Altocumulus altostratomutatus* (*Altocumulus* mutated from *Altostratus*)	*Cumulus* *Cumulonimbus*	*Cirrocumulus* *Altostratus* *Nimbostratus* *Stratocumulus*
Altostratus e.g. *Altostratus cumulonimbogenitus* (*Altostratus* formed from *Cumulonimbus*)	*Altocumulus* *Cumulonimbus*	*Cirrostratus* *Nimbostratus*
Nimbostratus e.g. *Nimbostratus cumulonimbogenitus* (*Nimbostratus* from *Cumulonimbus*)	*Cumulus* *Cumulonimbus*	*Altocumulus* *Altostratus* *Stratocumulus*
Stratocumulus e.g. *Stratocumulus nimbostratomutatus* (*Nimbostratus* that has completely mutated into *Stratocumulus*)	*Altostratus* *Nimbostratus* *Cumulus* *Cumulonimbus*	*Altocumulus* *Nimbostratus* *Stratus*
Stratus e.g. *Stratus fractus silvagenitus* (Broken wisps of cloud forming over a forest, due to high humidity near the tree canopy)	*Nimbostratus* *Cumulus* *Cumulonimbus* *Homo* *Silva* *Cataracta*	*Stratocumulus*
Cumulus e.g. *Cumulus cataractagenitus* (When the spray of a waterfall forms a cloud, in this case *Cumulus*)	*Altocumulus* *Stratocumulus* *Flamma* *Homo* *Cataracta*	*Stratocumulus* *Stratus*
Cumulonimbus e.g. *Cumulonimbus flammagenitus* (*Cumulonimbus* formed as a result of convection initiated by a wildfire or volcanic eruption. Also known colloquially as "pyrocumulus" or "pyrocumulonimbus")	*Altocumulus* *Altostratus* *Nimbostratus* *Stratocumulus* *Cumulus* *Flamma* *Homo*	*Cumulus*

Homogenitus
hogen

Homomutatus
homut

Flammagenitus
flgen

Silvagenitus
sigen

Cataractagenitus
cagen

FIVE SPECIAL MOTHER CLOUDS

Five particularly special mother clouds were only accepted as legitimate clouds by the WMO in 2017 (when the most recent online edition of the *International Cloud Atlas* was published). Each of these clouds is unique because it grows and develops as a consequence of specific natural or human "mother" factors, which are usually localized.

Homogenitus
Any cloud that owes its origin directly to human activity is given the appendage *homogenitus*, for example a freshly formed airplane condensation trail (contrail) *Cirrus radiatus homogenitus*. Similarly, *Cumulus* clouds that form above the warm industrial plume of a power station are referred to as *Cumulus homogenitus*.

Homomutatus
Any cloud that has transformed or mutated completely from an earlier mother cloud that was anthropogenic in origin is given the suffix *homomutatus*. Airplane contrails that grow and spread out across the sky, transforming themselves completely into *Cirrostratus*, would be termed *Cirrostratus homomutatus*.

Flammagenitus
Natural forest fires, wildfires, or even an erupting volcano can initiate powerful convection, giving rise to towering *Cumulus* or *Cumulonimbus*. When this happens, they are given the additional mother name of *flammagenitus*. Convective updrafts in *flammagenitus* can be extremely violent, leading to powerful storm systems and extraordinary electrical storms. *Flammagenitus* clouds are also commonly referred to as "pyrocumulus" or "pyrocumulonimbus."

Silvagenitus
If you live near a woodland or a large forest, especially one dominated by conifer trees, you may occasionally notice broken wisps of cloud (*Stratus fractus*) forming over the treetop canopy, especially just after a heavy shower or a spell of rain. These unique cloud wisps are caused by high humidity at canopy level due to increased evaporation and evapotranspiration from wet leaves and conifer needles, and are given the additional mother classification of *silvagenitus*.

Cataractagenitus
When the spray of large waterfalls is broken up by the wind to form a local *Cumulus* or *Stratus* cloud, the cloud is given the additional mother name *cataractagenitus*. There is usually a *Cumulus cataractagenitus* present in the vicinity of Niagara Falls.

1.

2.

3.

1. ***Upper Fall of the Reichenbach: Rainbow* by J.M.W. Turner, 1810**
   With the Reinbach waterfall, in Switzerland, dropping over more than 800 feet (250 m), significant amounts of spray are likely produced, as evidenced by the rainbow. Were a cloud of spray to waft away upward, it would be classified as *Cumulus cataractagenitus*. It must have been late in the morning, or well into late spring/summer, when Turner painted the scene, as the rainbow arc hardly reaches above the horizon.

2. ***The Great Horseshoe Fall, Niagara* by Alvan Fisher, 1820**
   Depending on the time of year and the water discharge over the Falls, a *Cumulus* or *Stratus cataractagenitus* cloud usually develops in the vicinity of Niagara Falls, drifting away slowly. The best time to see it is during intensely cold weather in winter, when the freezing air is unable to hold on to any additional water vapor.

3. ***Murton Colliery* by John Wilson Carmichael, 1843**
   Nearly two centuries before it was inaugurated as an official mother cloud, *Cumulus homogenitus* was first depicted by artists such as Carmichael. Here, the contribution of both aerosols (polluting particles and gases) as well as water vapor to the atmosphere from the smokestacks is probably leading to the growth of the *Cumulus* clouds overhead, although other clouds already present nearby are noted.

4. ***Cotopaxi* by Frederic Edwin Church, 1862 (overleaf)**
   A relatively minor eruption plume is blowing downwind, with ash and smoke quickly settling on the ground below. The small cloud generated is therefore *Cumulus flammagenitus*. In contrast, much larger and highly explosive volcanic eruptions can reach well into the stratosphere. On these occasions and owing to the extreme heat released, the resulting "pyrocumulonimbus" clouds can grow into huge thunderstorms of extraordinary violence, yielding frenzied electrical activity, large hail, and destructive winds.

4.

Asperitas
asp

Fluctus
flu

Volutus
vol

Cavum
cav

FOUR MORE NEW CLOUDS

As well as five new mother clouds (page 22), seven more new special clouds—one new species (*volutus*), five new supplementary features (*asperitas, fluctus, cavum, cauda,* and *murus*), and one new accessory cloud (*flumen*)—were inducted into the WMO's latest edition of the *International Cloud Atlas*. The first four of these special clouds are discussed below.

Asperitas
With pronounced wave-like features forming on its underside, *asperitas* was finally recognized as being a unique cloud supplementary feature in the *International Cloud Atlas* in 2017. Often forming near the outflow region of active fronts or severe thunderstorms, the characteristic chaotic underside of *asperitas* is sometimes referred to as looking like the "inside of a whale's mouth," or to viewing a roughened water surface from below. *Asperitas* is distinct from the *undulatus* variety due to its chaotic character; when viewed using timelapse photography, the waves oscillate up and down considerably and migrate through the cloud, unlike *undulatus* (whose waves move with the cloud) or *lenticularis* (whose waves are geostationary).

Fluctus
Again, unlike the species *lenticularis* or the variety *undulatus*, which have long lifetimes and are both associated with stable and laminar airflows, *fluctus* refers to a particular type of brief, transient, unstable wave, known in physics as a Kelvin–Helmholtz wave. They are short-lived, lasting only a couple of minutes, and form as curly wave crests on the top surface of a cloud layer that may topple over and break, just like ocean waves. They are caused by strong winds blowing near the top surface of the cloud layer, which act to amplify any small disturbances into larger waves, before tumbling them over. If you see one, try to enjoy it while it lasts, because by the time you reach for your cellphone/camera, it will likely have dissipated and you will have missed it!

Volutus
In appearance, *volutus* is similar to the supplementary cloud feature *Arcus* (page 202): however, it is a "soliton" (solitary wave) that, unlike *Arcus*, is not attached to other clouds, hence it is designated

5.

as a separate species. *Volutus* is rarely seen and is found only in specific parts of the world during the humid season, such as the "morning glory" cloud that occurs near the Gulf of Carpentaria in northeastern Australia. This example is thought to be formed by the collision of the previous day's sea-breeze fronts over the Cape York peninsula.

Cavum

Cavum isn't a cloud at all—instead, it is a hole in a supercooled cloud that is produced by airplanes flying through it (the cloud droplets have a temperature well below freezing, but remain as liquid; see page 68). The airplane's exhaust particles seed the cloud with tiny ice nuclei conducive to making the cloud freeze. Once ice crystals are formed, a chain reaction is initiated, allowing the ice crystals in the cloud to grow at the expense of the water droplets, due to the Wegener–Bergeron–Findeisen process (page 69). Not long afterward, within ten or twenty minutes, the ice crystals will have grown large enough to fall out of the cloud as snowflakes, leaving a hole in the cloud (*cavum*) above, and a trail of icy precipitation (*virga*) below the hole.

Cavum are usually concentric in shape, but if the airplane flies at a constant altitude slicing through a continuous layer of cloud, it may leave a stretched or elongated hole of cloud-free conditions, a phenomenon colloquially known as a "distrail" (the opposite of contrail).

5. ***Sea Shore in Moonlight* by Caspar David Friedrich, 1835**
A rare capture of *asperitas* on canvas, more especially so as it also predates its acceptance by the WMO as an official cloud feature by nearly two centuries. Best viewed today using timelapse photography, *asperitas* oscillates and undulates wildly.

6. ***Wheat Field with Cypresses***
by Vincent van Gogh, 1889
With aspects of *asperitas*, *mamma*, and *fluctus* (all unnamed and unclassified clouds in 1889), it's almost as though van Gogh could "see" the wind and all its turbulent motions. That's essentially what *asperitas* is: a churning sea of waves, eddies, and whorls, cascading across the sky. Even the wheat stalks and cypresses swing and sway in resonance with the wind, as if they are waving. Altogether, it's the ultimate capture, and a perfect natural metaphor for van Gogh's state of health at the time.

Arcus
arc

UNIQUE STORM CLOUDS: *ARCUS*

The dramatic cloud born of the mighty *Cumulonimbus* is called *arcus* ("arch" in Latin). It occurs on what is known as the "gust-front," which arises and races outward from the base of a strong thunderstorm. In its most severe incarnation, *arcus* can unleash weather that has the potential to cause widespread damage, injury, and occasionally even death, although for the most part the weather associated with it is not as severe nor as violent as that of a tornado.

When freshly formed from a rapidly developing, violent thunderstorm, the dramatic arrival of *arcus* is almost worthy of a Hollywood movie scene. It all happens within a few minutes—on an otherwise warm, fine summer's day, the sky begins to darken, with thunder rumbling in the distance. Suddenly, a particularly angry, low-level, turbulent arch of cloud sweeps across the sky, accompanied by violent wind gusts, sending people scurrying for cover. Air temperatures plummet by 18°F (10°C) or more within seconds, and torrential rain, and occasionally large hailstones, lash down in biblical proportions, accompanied by frequent lightning and ear-splitting thunder. The wind may continue to increase, felling branches or whole trees, while lifting other debris high into the air.

What causes the gust-front and *arcus*?

Arcus is formed by cold, dense air rushing out from precipitation downdrafts near the base of a *Cumulonimbus*; the air has been cooled by large quantities of precipitation evaporating during its descent to Earth, and has therefore been made denser and heavier than the surrounding air, the negative buoyancy now forcing it downward. This happens because the process of evaporation uses up a large amount of energy, which is provided by the air itself (page 51). When this happens, only the largest raindrops and hailstones, which fall fastest, make it to the surface, dragging the cold air along with them.

As the cold pool hits the ground, in what meteorologists term a "microburst," it spreads out, more or less radially, from the base of the storm in the form of a density current. The arch of *arcus* itself forms in the moist air on the leading edge of this concentric pool of cold air racing outward from the storm's center, helped by the slight upward motion on the leading edge of the density current which "tucks" under the warmer air surrounding the storm, lifting it and nudging it

7.

7. ***July Thundercloud in the Val d'Aosta* by John Ruskin, 1858** Ruskin's *Thundercloud* should be understood as an experience spanning several hours, rather than being an exact meteorological snapshot. The high intensity and "emotional atmosphere" of the storm is well captured, incorporating most of the features expected from a powerful thunderstorm: rapidly building cumuliform turrets, ascending slantwise (right, center), the gust-front *arcus* (left, center) sweeping down the valley; intense streaks of precipitation (*virga*, top background) with the sky brightening beyond (an effect caused by large raindrops).

slightly upward. Similarly, cold air outflow from large *Cumulonimbus* thunderstorms often creates new instances of convection, known as "daughter cells," allowing the storm system to propagate forward and continuously rejuvenate itself at the same time.

In the mayhem created by the arrival of a powerful gust-front over urban and populated areas, the upward motion on its leading edge can lift outdoor furniture, garden ornaments, tents, and even marquees high into the air, which is one reason why the impacts of a severe gust-front can be easily mistaken for a mini-tornado. In this case, the winds are generally "straight-line," but are still powerful enough to fell whole trees, and generally interrupt and lay waste to most outdoor activities, including sporting events, markets, and music festivals. Every year, there are injuries and deaths that arise from falling trees and flying debris caused by the sudden arrival of severe gust-fronts associated with *arcus*.

The green light

The next time *arcus* comes your way, it is worth looking out for an eerie dark green hue emanating from under the emerging arch, which becomes visible to the naked eye for a short while just before the gust-front arrives. Legend has it that if you see it, a tornado or large hailstorm, or both, is about to hit. Fortunately, there is no scientific evidence of any such relationship; the dark green color is instead attributed to the combination of deep clouds containing a large volume of precipitation, acting together with a particular alignment of the Sun's rays to give off a dark green hue.

Arcus, when made manifest, is exactly what it sounds like: a dark, dramatic, threatening, awe-inspiring, fast-moving arched roll of cloud, the arch itself being an artifact of both perspective and the radial curvature of the cloud. Although its associated gust-front is generally a nuisance and occasionally lethal, it is not as deadly as a tornado, but it is far more common.

Murus
mur

NOT TO BE CONFUSED WITH…

Wall clouds are not the same as the fast-approaching curtain of cloud of an impending squall, which sometimes takes on the appearance of a wall of cloud rushing in, but instead is a dense sheet of *virga* caused by heavy precipitation, sometimes accompanied by *Arcus* (page 202). Squalls can also produce violent weather matching that of supercells, but the winds are usually of a straight-line nature, rather than being tornadic.

Murus should also not be confused with the "eyewall" cloud of a hurricane—which is a completely different, but even more destructive, meteorological phenomenon. At first glance, waterfall föhn clouds (a type of *lenticularis* mountain cap cloud) can also look like stationary wall clouds, but they are an indicator of a stable atmospheric environment: the opposite of *Cumulonimbus*.

MURUS

Murus, *cauda*, and *flumen* are rare clouds associated only with a distinct type of mature, severe, and powerful *Cumulonimbus* known as a "multicell" or "supercell" thunderstorm. These rare clouds are usually restricted to certain latitudes and times of the year, namely continental areas in the warm season—and particularly over North America.

How *murus* develops

Murus is more commonly known by storm chasers as a "wall cloud"—hence the name, which means "wall" in Latin. It is a prominent, threatening, and persistent lowering of the base of a *Cumulonimbus* (page 98) just below the center of a severe multicell or supercell thunderstorm.

Wall clouds form directly underneath the strongest updrafts of the storm, so the region is usually free of rain or hail; the storm's precipitation falls elsewhere, tilted away from the storm updraft by strong winds at mid- and upper levels of the *Cumulonimbus*. Rotation of *murus*, mostly counterclockwise in the northern hemisphere (and vice versa in the southern hemisphere), is common in severe storms and is usually the precursor of a tornado (page 208).

As they form at the lowest part of a storm structure, if you want to spot a *murus*, your view of the storm must be unrestricted by topography, trees or nearby buildings; this is one reason why the Great Plains or prairies of North America are good places to view them.

The cloud lowering is caused by air converging beneath the storm's center. This is aided by reduced air pressure beneath the warm updraft—effectively attracting the converging air—and increased amounts of atmospheric moisture in the vicinity of the storm, due to both the inflow of moisture and the evaporation of falling precipitation nearby. *Cauda* and *flumen* are also storm inflow clouds, shaped like beavers' tails.

Murus may persist as an identifiable feature beneath a large *Cumulonimbus* for up to 10 minutes or more, before decaying into scraggy patches of *pannus* (scud) or re-forming as another fresh wall cloud nearby. Less often, *murus* may suddenly descend even further, forming a spout or funnel cloud (*tuba*, page 206), or a tornado if it touches down on the ground.

THE EYEBROW CLOUD: *SUPERCILIUM*

Supercilium

Have you ever looked upward in awe at a mid- or high-level cloud formation, perhaps raising an eyebrow in surprise? Well, if it is the beautiful *supercilium* that has caught your eye, it may be the cloud itself that is raising an eyebrow back at you, for *supercilium* is colloquially known by the cloud-spotting fraternity as "the eyebrow cloud," *supercilium* being Latin for eyebrow.

Occurring rarely, with a temporal frequency equivalent to that of *fluctus* (page 198), *supercilium* appears only briefly as a wafer thin, wispy, irregularly distributed, and wavy cloud, and is found only in turbulent airflows over steep or jagged mountain peaks, usually coinciding with strong mountain-level winds. Unlike the smooth lens shape of the more common mountain species *lenticularis*, which has streamlined and laminar profile, the air stream passing through *supercilium* appears to be tumbling over and breaking in a turbulent way across a range of scales, both passing through, and advecting the cloud forward at the same time. However, in common with *lenticularis*, the cloud sometimes produces iridescence, meaning its droplets, like *lenticularis*, are likely to be tiny but regularly sized.

Supercilium also shares some attributes with the cloud varieties *undulatus*, *duplicatus*, *lacunosus*, and *fluctus*, but lacks any of their geometrical regularities. For example, similar to *pile d'assiettes* (page 198), it may consist of several discrete layers, stacked vertically upon one another: however, unlike *pile d'assiettes*, these break or tumble in a turbulent manner, almost on the microscale.

At the time of writing, *supercilium* is not listed as an official cloud in the WMO's *International Cloud Atlas*. However, fresh evidence of its occurrence is being constantly gathered and amalgamated by global cloud-spotting networks.

Supercilium mostly occurs at *Altocumulus* levels but may also be spotted in association with both *Cirrocumulus* and *Stratocumulus*.

Tuba
tub

TUBA

Tuba is the official supplementary cloud name for the visible manifestation of either a funnel cloud or a tornado—the latter if the vortex reaches Earth's surface. A *tuba* must be connected to the base of a *Cumulonimbus* (or less commonly, a *Cumulus congestus*); this means minor dust devils, or any of their close relatives such as mini-whirlwinds, "hay devils," and "snow devils" that might occur on otherwise fine days, are neither tornadoes nor *tuba*.

Tornadoes are a serious business, posing a major threat to life and property when they occur; however, the good news is that most *tuba* are weak and short-lived. The big, dramatic, awe-inspiring twisters featured online or in Hollywood movies are rare, and they do not come directly for you, as the movies usually depict. Indeed, they are deliberately sought out by extreme weather enthusiasts chasing the thrill of the closest encounters, and ultimately the most "likes."

A large *tuba* typically evolves from very low, ragged "scud" clouds (*pannus*) that are sometimes found in the vicinity of a storm's wall cloud (*murus*; page 204) when strong rotation that already exists around the wall cloud gets concentrated into a single column of spinning air. At this stage, the newly formed *tuba* usually takes on the shape of a pronounced, vertically oriented cone or funnel: this initial developmental stage can happen within a matter of a few tens of seconds to a few minutes. In the most severe storms, this cone may descend completely to Earth's surface, then widen and stretch out across the ground for up to 1 mile (1.6 km) in width, forming what is called a "wedge tornado"—the most severe, dangerous, and long-lasting of all twisters. More commonly, however, after touchdown, the vortex of a mature tornado will eventually start to tilt, stretch, and lengthen considerably, before twisting and contorting itself in the sky like a snake or a loosened rope. Once this happens, it signals that the most severe stage of the tornado has passed—although it can still be dangerous—and the tornado will likely dissipate sometime over the following few minutes.

Dangerous tornadoes

Tuba refers only to the cloudy part of a funnel or a tornado, the zone that is saturated and has condensed into a cloud. In a humid air mass, this happens around a tornado funnel due to the lower air pressure within the vortex—the low pressure arises due to the expansion of air, which causes cooling, and this in turn causes condensation. In drier environments, for example in desert storms where the *Cumulonimbus* storm clouds have high bases, tornadoes that touch down may only become visible due to the cloud of surface dust and debris that they generate at ground level. In these cases, only a very small *tuba* may be identifiable near the cloud base, with most of the tornado occurring in largely clear air. Such tornadoes can still be extremely dangerous, particularly because they are largely invisible.

8. ***Tornado over St. Paul***
by Julius Holm, 1893
Holm reportedly painted this scene from an early photograph reproduced on a souvenir card, commemorating the deadly twister of July 13, 1890. From this elevated vantage position, we can clearly see the turbulent base of the *Cumulonimbus* (dark cloud, top), the lowered wall cloud (*murus*, lighter shade, center) with the tornado (*tuba*) protruding prominently from its base. The twister killed six people.

8.

In the United States, tornadoes kill on average about 75 people per year, but this varies considerably from year to year. For example, a single tornado on March 18, 1925—the infamous "Tri-State Tornado"—killed 695 people; in 1910, however, the annual nationwide death tally was 12. More recently, in 2011, there was a sudden spike in death statistics, with numbers reaching 553. Overall, however, there was a clear downward trend in the twentieth century in the number of deaths caused by tornadoes in the United States, by the order of around 50 percent. This is not due to fewer storms, but more likely due to improved weather forecasts, increased social resilience, and better public awareness of what to do (and what not to do) in the event of an approaching tornado. The annual US death toll from tornadoes is only a tiny fraction of the total number of worldwide deaths arising from severe weather each year (approximately 20,000).

The terminology surrounding tornadoes differs from place to place, as well as changing through time. For example, according to the Google Books Ngram Viewer, the Middle English–derived term "twister" was most used in American English publications around 1900 but has since reduced in frequency—but not in British English. The use of the terms "tempest" and "whirlwind" has dramatically reduced since the late nineteenth century; today, a whirlwind refers only to an inconvenient, but hardly deadly, dust-devil, or similar nuisance. Over water, weak to moderate tornadoes are termed "waterspouts," while over land they are referred to as "landspouts," but again usage of these two terms is erratic through time and place.

Nacreous clouds

HARBINGERS OF DOOM: NACREOUS CLOUDS

Also known as "mother-of-pearl clouds" or "polar stratospheric clouds," nacreous clouds are not your normal, everyday type of cloud. Found only above the Arctic, Antarctic, or at high latitudes, nacreous clouds form in the stratosphere. This atmospheric layer lies above the troposphere where all the everyday weather-producing clouds are found.

In the stratosphere, at an altitude of between 12 and 25 miles (20 and 40 km), conditions are effectively alien; the air is extremely tenuous and thin, with pressures ranging from one-fifth to one-hundredth of its value at sea level. This, together with relatively high amounts of ozone (which is poisonous for humans when inhaled), extremely low air temperatures, and very dry conditions, would make it an impossible place for humans to survive. However, such extreme conditions permit the formation of nacreous clouds, which are astonishingly beautiful and extraordinary in many ways.

How nacreous clouds form

The striking pearlescent appearance of nacreous clouds requires an air temperature of at least -108°F (-78°C) for their formation. Such low temperatures are generally found only above Antarctica, and to a lesser degree over the Arctic, and only during their respective winters. But very occasionally during the northern winter, when the stratospheric polar vortex shifts slightly off-center, cold stratospheric air may filter down toward lower latitudes from the Arctic.

When this happens, and is coupled with pronounced mountain wave activity (page 156), nacreous clouds may be spotted in the cold crests of atmospheric waves to the lee of the Alaskan, Canadian, and Scandinavian mountains. Very rarely, they may also be seen over the smaller hills and mountains of Scotland, Ireland, and northern England.

9. ***The Scream* by Edvard Munch, 1893**
Munch was apparently out for a walk after sunset and saw "the clouds turn blood red" (a final phase of the coloration of nacreous clouds just before they are eclipsed). Originally called "The Scream of Nature," a recent series of papers in the journal *Weather* (published by the Royal Meteorological Society) argue that the painting is an emotional reaction to the sighting of the nacreous clouds, the waves and hues of which are portrayed in an Impressionist fashion in the background (top).

Mountain cap cloud

Banner cloud

Föhn wall

SOME SPECIAL OROGRAPHIC CLOUDS

There are other special "orographic" clouds that are much rarer than *lenticularis*. This is because they remain closely anchored to an individual mountain peak and do not migrate far downstream. Here, we take a look at three of them.

Mountain cap cloud
When a smooth and stable lenticular cloud forms directly over the slopes or peak of a mountain, shrouding it in a "skirt" of cloud, sometimes with the peak emerging from the top, it is colloquially known as a "mountain cap cloud." The best place to spot these are over isolated mountain peaks, such as volcanoes, during spells of humid and stable weather.

Banner cloud
During strong mountain summit winds, a banner cloud may form immediately to the lee of a steep, pyramidal, mountain peak. It appears as wisps or a plume of ragged cloud (*Stratus* or *Cumulus fractus*) that appears to stream away from the summit, before dissipating a short distance downwind and re-forming quickly again close to the mountain summit. Its formation is thought to be due to a slight upslope wind on the leeward edge of the mountain peak that cools and saturates the ascending air, induced by the stronger winds blowing around and over the windward mountainside. Like *lenticularis*, the banner cloud is geostationary. In fine weather, a banner cloud can commonly be seen directly to the lee of the Matterhorn in Switzerland.

The Föhn wall
The *Föhn* is a warm, powerful wind that descends the northern slope of the Alps during southerly airflows, or the southern slopes during a northerly wind. The cloud itself, which can form over the edge of any mountain range in suitable weather, is dramatic and appears like a smooth wall of cloud lying on the summit peaks, with occasional tendrils reaching downslope. However, it is a sign of a warm and stable atmosphere, and conditions are likely to remain fine and dry at the viewing location, albeit with a risk of strong winds. The "tablecloth" of Table Mountain, Cape Town, South Africa, is a well-known Föhn wall cloud.

10.

10. ***Norsk fjordlandskap med regnbue*** **by Andreas Achenbach, 1839**
Mountains make clouds—sometimes excessively so—and in exposed maritime environments where the hills rise steeply from the shore, such as here in western Norway, cloudy rainy weather is very much the norm. Driven by strong westerly or southwesterly winds, the airmasses arriving from across the ocean are usually laden with moisture. As soon as the air impinges upon the first coastal peaks, it is forced upward forming clouds, releasing latent heat and, not long afterward, huge quantities of precipitation.

Th. Kittelsen

11. ***Far, Far Away Soria Moria Palace Shimmered Like Gold***
by Theodor Kittelsen, 1900
Soria Moria is a popular Norwegian fairytale, but Kittelsen largely adheres to meteorological idealism in his portrayal of the clouds surrounding the distant fantasy land; the scene is not unlike one which we might witness today from an airplane flying between two decks of cloud. The lowest cloud layer is a mixture of orographic *Stratus* and *Stratocumulus lenticularis*, mimicking the contours of the hills over which it flows. The upper cloud deck (also stratiform) is somewhat less realistic; one might expect to see instead a more lightly dappled layer of *Altocumulus*.

Noctilucent clouds

NIGHT CLOUDS: NOCTILUCENT

Noctilucent, or "night-shining," clouds are the highest of all clouds in the atmosphere, having a distinctive, brilliant, silvery-white or electric blue appearance. They form in the mesosphere at an altitude of 50–55 miles (80–90 km), where air pressure is 1/100,000th of that at sea level and air temperatures fall below -184°F (-120°C). We see them from the middle latitudes on clear nights during the early-to-mid-summer months of each hemisphere, close to the poleward horizon.

Why do they shine at night? At these times, the Sun lies only a few degrees below the horizon, and owing to their extreme altitude, the clouds remain brightly illuminated, in stark contrast to the darkening sky at ground level, providing ideal viewing conditions.

Although perhaps not quite as awe-inspiring and ethereal as nacreous clouds, they are nonetheless a spectacular sight to behold. When imaged used timelapse photography, they reveal themselves as tenuous, or very thin, wavy threads or "eyebrows" moving in a rippling motion reminiscent of breaking waves, with their overall form not dissimilar to that of *Cirrocumulus undulatus* or *lacunosus*, or even *supercilium* (page 205).

Out of this world

Each noctilucent cloud is made up of billions of tiny ice crystals. Their magnificent blue color is thought to be caused by the absorption by ozone of the illuminating light.

Noctilucent clouds lie far above the clouds that produce our weather, which are nearly all found in the troposphere below an altitude of 9 miles (15 km); it is therefore unlikely that water vapor or cloud nuclei can rise undetected from the troposphere to the mesosphere to form these clouds. Indeed, recent data from NASA indicates that the clouds form on minute particles of meteor or cosmic dust, either originating from the solar system, or beyond it—they really are out of this world!

Another harbinger?

Noctilucent clouds occur only in the summer months when the mesosphere is coldest. This fact may seem counterintuitive at first but becomes more obvious when we consider that the troposphere, which is closest to Earth's warm surface, heats up in

12.

the summertime, leading to its expansion. This expansion then "lifts" the layers above it slightly, including the mesosphere, cooling them adiabatically (page 48), which in turn allows the threshold low temperature for their formation to be surpassed.

This may explain why reports of noctilucent cloud are becoming more frequent. There are no known observations of them before the late nineteenth century, whereas in recent years we have seen some of the most widespread displays and earliest dates of observation ever observed.

Much like their destructive stratospheric (nacreous) cousin, it seems that the expansion of the globally warmed troposphere, and the consequent "lifting" of the stratosphere and mesosphere above it, may be responsible for the increased sightings. Noctilucent clouds may therefore also be another beautiful but equally terrible harbinger of the future (page 208).

12. ***The Starry Night* by Vincent van Gogh, 1889**
Although atmospheric turbulence is a chaotic phenomenon, there is a known rate at which eddies are produced and dissipated in the air. This is known as the "turbulent cascade" and was discovered by the Russian scientist Andrey Kolmogorov in 1941. In a recent research study on the sizes of the whirls depicted in *The Starry Night*, it was discovered that they closely match the pattern predicted by Kolmogorov. So it seems van Gogh knew advanced theoretical physics, as if by instinct, long before Kolmogorov did!

GLOSSARY

Actinoform (clouds): Large star-shaped clusters of open-celled *Stratocumulus* spanning some 60–180 miles (100–300 km), mainly found over the Pacific Ocean. Only observable from satellites in space.

Adiabatic cooling/warming: The automatic cooling of air when it expands, or warming when it is compressed, without any net heat exchange with surrounding ambient air. For example, a dry air parcel lifted from sea level to 1,000 feet (305 m) will cool by 5.4°F (3.0°C) due to expansion; the reverse process of air descending by 1,000 feet leads to a net warming of the same amount, 5.4°F.

Advection: The horizontal movement and transfer of weather systems, from a fixed observer's point of view on Earth.

Aerosol: Minuscule (can be sub-microscopic) particles of solid or liquid suspended in the air. See also **cloud condensation nuclei**.

Airmass: A large body of air with relatively similar temperature and humidity. Air masses may develop over a wide ocean area, or over a continental land mass, before being advected (moved) elsewhere.

Albedo: The reflectivity of a surface or substance, usually expressed as a percentage or fraction of the incoming radiation. For example, the albedo of fresh snow in visible light is about 90 percent (or 0.9).

Anabatic wind: An upslope breeze in mountainous area, as part of the daytime mountain-valley air circulation system (page 101). From the Greek *anabatikos*, "the person who ascends." See also **katabatic wind**.

Anthropocene: A term originally coined by atmospheric chemist Paul Crutzen in the early 2000s to describe "the age of humans" as a new geological epoch.

Anticyclone: A high-pressure weather system bringing settled conditions, extending over a wide area covering hundreds to thousands of miles. Anticyclones spin clockwise in the northern hemisphere, vice versa in the southern hemisphere. See also **cyclone** and **synoptic scale**.

Anvil: The glaciated crown or "flat-top haircut" of a *Cumulonimbus capillatus*. Designated officially as the supplementary feature *incus*. Named after the shape of the blacksmith's anvil.

Back (wind): Wind direction moving counterclockwise (anticlockwise) around the compass. For example, a southwesterly wind is said to "back" when it changes to a southerly wind.

Billows: Attractive, white or brightly lit *undulatus* variety of either *Altocumulus* and *Cirrocumulus*, where the wave is heaped but not overhanging or breaking like *fluctus*.

Boundary layer: The layer of air closest to Earth's surface, in which frictional effects are observed, with a typical vertical thickness of 1,000–3,000 feet (300–1,000 m).

Buoyancy (meteorology): The tendency for air parcels to rise of their own volition until they attain a level of equal density with the surrounding air. Buoyancy in the atmosphere is a direct consequence of Archimedes' Principle. Such air is said to be "unstable." Buoyancy, or instability, is the opposite of stability.

Cloud condensation nuclei (CCN): Tiny suspended aerosols in the air which are hygroscopic. They act as nuclei for the formation of cloud droplets. See also **aerosol**.

Cloud seeding: Artificial seeding of clouds with chemicals such as silver iodide or sodium chloride, usually by airplane or fired rockets, with a view to increase or alter precipitation at ground level.

Cloud streets: Parallel rows of convective cloud, *Cumulus* or *Stratocumulus* (variety *radiatus*), which align themselves with the prevailing wind direction. Each row consists of a large cylindrical eddy (or "roll") of air, circulating in the opposite sense to its neighbors on either side. After initial formation, the horizontal distance between individual cloud streets widens downwind as the size of each eddy increases within the boundary layer.

Cold front: The moving boundary between warm and moist, and colder and drier airmasses, often bringing thick clouds (*Nimbostratus* or *Cumulonimbus*) and precipitation, and where the colder air follows the passage of the front.

Convergence: The meeting, or coming together, of air, as if the winds from around the compass are blowing toward a central zone. If occurring at the surface, the net result is the ascent of air. Conversely, convergence near the tropopause will usually cause descent. See also **divergence**.

Corona(e): Colored "mother-of-pearl" rings that appear around the Sun or Moon, caused by the diffraction of light by small particles (usually cloud droplets) in the atmosphere. See also **iridescence**.

Cyclone: A low-pressure weather system than spins counterclockwise (anticlockwise) in the northern hemisphere, vice versa in the southern hemisphere, and extending over a wide area typically covering hundreds to thousands of miles. See also **anticyclone** and **synoptic scale**.

Dalton's Law: The total pressure exerted by a mixture of gases, such as air, is equal to the sum of their partial pressures.

Density current (meteorology): When a lobe of cool, dense air undercuts or intrudes under a zone of much warmer and less dense air, maintaining its cohesion in the process. Occasionally, sea-breezes, *volutus*, and *arcus* provide examples of atmospheric density currents, in the form of solitons or undular bores.

Diffraction: The apparent bending or spreading out of light waves as they pass around a tiny obstruction, or around a sharp edge, causing colored interference patterns. When tiny cloud droplets or ice crystals are of the same order of magnitude as the wavelength of light, diffraction dominates over refraction. See also **iridescence** and **corona(e)**.

Diffusion (gas): The random movement of gas molecules which therefore facilitates its spread from regions of high concentration to areas of lower concentration.

Divergence: The spreading out or "stretching" of air, as if the winds are blowing away from central zone. When divergence occurs at the surface, the air may have just descended prior to diverging. Conversely, divergence at upper tropospheric levels can lead to lowering of air pressure and the ascent of air from below.

Downdraft: A coherent area or current of air moving downward in a cloud, usually a *Cumulonimbus*. A powerful or violent downdraft may also produce a "microburst" at ground level, as well as *arcus* or *volutus* clouds on its leading edge as it spreads outward from the storm center.

Dynamical meteorology: Refers to the study of atmospheric science by means of the fundamental equations of motion, thermodynamics, and radiation. "Dynamics" in a more general sense refers to properties of synoptic scale weather systems and the motion of air.

Glaciation: The freezing of clouds.

Gravity wave (meteorology): An atmospheric wave causing air to oscillate vertically, in a similar way to water waves, as gravity tries to restore hydrostatic balance (page 52) to the atmosphere. Gravity waves in the atmosphere can be transient (such as in *undulatus*, *asperitas*, *fluctus*, *supercilium*, or *volutus*) or geostationary (*lenticularis*).

Habits (meteorology): The flat faces and external structure of ice columns and hexagonal ice crystals, which give rise to the reflection and refraction of incoming light rays. We may see these as ice haloes when viewed from a distance.

Hadley Cell: Quasi-permanent hemispheric-scale features of the subtropics, in which upper air originating in the convective *Cumulonimbus* cells of the tropics moves poleward and begins to sink at latitudes of around 25–30°N/S, before it returns on its journey equatorward as the Trade Winds.

Hydrometeor: A comprehensive term referring to any particle of liquid water or ice in the atmosphere, for example cloud droplets, fog, mist, drizzle, raindrops, hail, sleet, snow, and other ice crystals.

Hygroscopic: The property of a substance (such as an aerosol or cloud condensation nuclei) to attract and condense water vapor on its surface before saturation is reached. For example, water vapor may condense on large seasalt aerosols at relative humidities of 78 percent and above.

Infrared light: The part of the electromagnetic spectrum between 700 nanometers (0.7 microns) to 1 millimeter (1,000 microns). Infrared radiation is the dominant radiation given off by objects at terrestrial temperatures. Humans are unable to see it.

Inversion: When the normal rate of drop in air temperature with height is reversed, i.e., the air temperature increases with height. Inversions are usually associated with stable weather conditions, meaning that any rising air which encounters an inversion will tend to sink back down to its original position.

Iridescence: See page 78.

Katabatic wind: A downslope airflow or gentle breeze in a mountainous area, as part of a nocturnal mountain-valley air circulation system. From the Greek *katabatikos*, "to descend." See also **anabatic wind**.

Kelvin (K): The absolute scale of temperature, which starts at the lowest possible temperature of 0K, equivalent to -273.15°C on the Celsius scale. The Kelvin scale has the same magnitude as the Celsius scale, whereby 1K=1°C. Named after Lord Kelvin.

Lapse rate: The rate of change of air temperature with height.

Latent heat (meteorology): The "hidden" energy released by water vapor upon condensation amounting to about 590 calories per gram of water (or 2,500 kilojoules per kilogram) at terrestrial temperatures, which, in a cloud, heats the water droplet itself. Latent heat is also released upon freezing (80 kcal, or 334 kJ/kg). The same energies are taken from the surrounding environment during evaporation, or during melting, respectively.

Low-pressure weather systems: See **cyclones**.

Mesosphere: The part of Earth's atmosphere between 30–50 miles (50–80km) in altitude.

Microburst: A large downdraft from a severe *Cumulonimbus*, which spreads violently outward when reaching ground level, bringing high winds and torrential precipitation. See also **downdraft**.

Micron: One-thousandth of a millimeter (0.001 mm), or 1×10^{-6} of a meter, also called a micrometer, and often denoted as µm.

Mother-of-pearl (clouds): Another common name for nacreous clouds or polar stratospheric clouds (see page 208).

Mountain wave: A geostationary gravity wave in which *lenticularis* clouds form. See also **gravity wave**.

Multicell (thunderstorm): A severe type of thunderstorm where individual single storm cells merge into a larger, complex, and severe storm system. See also **supercell**.

Nacreous clouds: Known also colloquially as "mother-of-pearl clouds," their official name is polar stratospheric clouds. Found only in the Arctic, Antarctic, or at high latitude, nacreous clouds form in the stratosphere. See **mother-of-pearl (clouds)**.

Nowcast: A very short-term weather forecast spanning the next 0–6 hours.

Orography: Hills and mountains.

Ozone layer: A vital part of the stratosphere containing a high concentration of ozone, which absorbs most of the harmful ultraviolet radiation incoming from the Sun, and warming the stratosphere in the process.

Provenance (of clouds): The origin of a cloud. For example, clouds may modify, or mutate, from a "mother" cloud (see page 192).

Pyrocumulus/pyrocumulonimbus (cloud): A commonly used name for *Cumulus*/*Cumulonimbus flammagenitus* (see page 194).

Radiational warming/cooling: All objects with a temperature above absolute zero emit electromagnetic radiation. Energy gained from a warm source (e.g., the Sun; from a warm atmosphere) therefore leads to radiational warming. Energy lost to a cooler target (e.g., a clear sky at night) therefore causes radiational cooling. Earth and its envelope of clouds cool by emitting radiation mostly at infrared wavelengths, which is invisible to our eyes.

Relative humidity: The degree of saturation of air at the current air temperature, expressed as a percentage between 0 and 100. In general, fog and clouds form when the relative humidity reaches 100 percent, allowing the excess moisture in the air to be condensed.

Scattering (meteorology): When incident electromagnetic radiation, typically light, is reflected diffusely in all directions, rather than in a selected direction(s).

Shelf cloud: Another name for *arcus* (see page 202).

Snow grains: A conglomeration of ice crystals found in turbulent, precipitating clouds.

Specific humidity: The amount of gaseous water vapor in the air, expressed as a ratio of the mass of water vapor (in grams) per unit mass of air (in kilograms).

Stability: The tendency of air parcels to return to their initial position after being forced upward or downward. Stability is the opposite of instability. See also **buoyancy**.

Storm: A nonexclusive generic term referring to many types of severe weather, including thunderstorms, windstorms, middle latitude cyclones (lows, or depressions), tropical cyclones (tropical storms, typhoons, hurricanes, cyclones), squalls, polar lows, dust storms, haboobs, and derechos. A storm may bring thunder and lightning, heavy precipitation, or strong winds, or a combination of all three.

Stratopause: The boundary between the stratosphere and the mesosphere.

Stratosphere: The part of the atmosphere above the tropopause, containing the ozone layer at altitudes of approximately 9–31 miles (15–50 km).

Stratospheric polar vortex: A planetary-sized circumpolar air circulation system present in the lower stratosphere during fall, winter, and spring months, revolving around both poles at approximately 50–0° of latitude.

Sublimation (meteorology): The evaporation from water from its solid (ice) to gas phase, skipping the liquid stage.

Supercell: A very large, powerful and mature *Cumulonimbus* that has become a unique weather system in its own right. The characteristics of a supercell storm, which differentiate it from other *Cumulonimbus*, include large hail, a wall cloud (*murus*), a powerful, single, well-separated and self-regenerating updraft, violent downdrafts, and being a "right mover" (whereby the storm doesn't move directly downwind, but rather propagates to the right of steering wind at a sharp angle). Supercells are also more likely to produce severe tornadoes than other storms.

Synoptic scale: The broad-scale weather features of the atmosphere, such as high pressure (anticyclones) or low-pressure systems (cyclones), warm and cold fronts, operating over scales of hundreds to thousands of miles.

Terrestrial: Meaning "relating to Earth."

Topography: The shape and variability in altitude of the surface landscape. See also **orography**.

Trade winds: Prevailing surface winds encountered in the equatorward arm of a Hadley Cell, which usually have a northeasterly directional component in the northern hemisphere and southeasterly in the southern hemisphere.

Tropopause: The boundary between the troposphere and the stratosphere. It typically lies at about 4–5 miles (7–8 km) in altitude above polar regions, and 8–10 miles (12–16 km) above the tropics.

Troposphere: The lowest part of the atmosphere closest to Earth's surface, from 0–9 miles (0–15 km), where almost all weather and clouds occur. At its upper limit lies the tropopause; above it lies the stratosphere.

Twomey Effect: A mechanism which increases the reflection of incoming solar radiation by low-level clouds, through the addition of aerosols into the atmosphere, making the clouds brighter.

Undular bore (meteorology): A propagating wave or soliton such as *volutus* (see page 198). See also **density current**.

Updraft: A coherent area of air moving upward in a cloud, usually a *Cumulus* or *Cumulonimbus*.

Vapor pressure (meteorology): The partial pressure of water vapor in the atmosphere. See also **Dalton's Law**.

Veer (wind): Wind direction moving clockwise around the compass, for example the wind is said to "veer" when a southwesterly wind changes into a westerly one.

Wall cloud: Popular and common name for *murus* (see page 204).

Warm front: The moving boundary between cool (and usually drier) and warm (and usually moister) airmasses, often bringing thick clouds (*Nimbostratus*) and precipitation, and where the warmer air follows the passage of the front.

Wind shear: A change in the magnitude of the wind (its speed and/ or direction) with altitude.

WMO: The World Meteorological Organization, formed in 1951 from the International Meteorological Organization. It administers and facilitates international cooperation on meteorology and related issues.

INDEX

D

E

F

G

H

I

PICTURE CREDITS

Alamy Photo Library: Historica Graphica Collection/ Heritage Images 25; The History Collection 30; Zuri Swimmer 35; The Picture Art Collection 69; The Print Collector 79 B; Ashmolean Museum of Art and Archaeology/Heritage Images 105; Heritage Image Partnership Ltd 125; World History Archive 126–127 ; Ashmolean Museum of Art and Archaeology / Heritage Images 166–167; piemags/RTM 169 B; The Print Collector 183; classicpaintings 184–185.

Birmingham Museum and Art Gallery: Creative Commons 107; Google Art Project 120–121.

Bridgeman Images: 12–13; 81; Scott Polar Research Institute 79 T; Bury Art Museum & Sculpture Centre 82–83; Photo Josse 149 ; The Wilson 156; Science and Society Picture Library 161; Ashmolean Museum 169 T; Photo Josse 171; National Trust Photographic Library 177 B. Sunderland Museums 195 B.

The Cleveland Museum of Art: Gift of Mr. and Mrs. J. H. Wade 73.

Cooper Hewitt, Smithsonian Design Museum Gift of Louis P. Church: 11; 16–17; 50 B; 55; 99; 117 B; 119; 146–147; 174–175; 178–179.

Detroit Institute of Arts Museum: Founders Society Purchase, Robert H. Tannahill Foundation Fund, Gibbs–Williams Fund, Dexter M. Ferry Jr. Fund, Merrill Fund, Beatrice W. Rogers Fund, and Richard A. Manoogian Fund 196–197.

The J. Paul Getty Museum, Los Angeles: 131 B.

Hamburger Kunsthalle: 47; 191.

Indianapolis Museum of Art (IMA): 101.

Ivan Konstantinovich Aivazovsky: 111 T.

The Metropolitan Museum of Art: The Whitney Collection, Gift of Wheelock Whitney III, and Purchase, Gift of Mr. and Mrs. Charles S. McVeigh, by exchange 49; Morris K. Jesup Fund 60–61; 131 T; Bequest of Miss Adelaide Milton de Groot 93; H. O. Havemeyer Collection, Bequest of Mrs. H. O. Havemeyer 95 B; 181; Gift of Mrs. Leon L. Watters, in memory of Leon Laizer Watters 96 T; Robert Lehman Collection 96 B; Purchase, The Annenberg Foundation Gift 201.

Minneapolis Institute of Art Collection: The Ethel Morrison Van Derlip Fund 207.

Museum of Modern Art: Google Art Project 215.

Nationalmuseum Stockholm: 140–141; 145; 122–123; 209.

Nasjonalmuseet for kunst, arkitektur og design, The Fine Art Collections: 18–19; 95 T; 102–103; 117 T; 129; 133 T; 177 T; 211; 213.

The National Gallery: 95 M.

The National Gallery of Art: Chester Dale Collection 155 B.

The New Art Gallery Walsall: Garman Ryan Collection 1973.023.GR : 112–113.

Philadelphia Museum of Art: John G. Johnson Collection : 134 –135.

The Royal Society: 39 T.

Board of Trustees of the Science Museum: 31; 32; 39 B, 173.

Smithsonian American Art Museum: Gift of William T. Evan 90–91; 195 M.

Wellcome Collection: 45.

Wikimedia Commons: National Portrait Gallery 30; Nationalmuseum 181; 122–123 181; Hamburger Kunsthalle 199; 203.

Wilson, E., Wilson, D.M. and Wilson, C.J., 2011. *Edward Wilson's Antarctic Notebooks*. Reardon Pub: 186; 187.

Yale Center for British Art, Paul Mellon Collection: 4–5; 7; 8–9; 20–21; 40–41; 42–43; 50 T; 63; 66–67; 76–77; 84–85; 89; 109; 111 B; Given in honor of Patrick Noon, Curator of Prints, Drawings & Rare Books (1979–97), from the collection of Iola S. Haverstick : 107 B; 115; 133 B; 136–137; 143; 151; 153; 155 T; 162–163; 190–191; 195 T;